ETHNOBOTANY

'PEOPLE AND PLANTS' CONSERVATION MANUALS

Manual Series Editor:
Martin Walters

Manual Series Originator:
Alan Hamilton
Plants Conservation Officer
WWF International

'People and Plants' is a joint initiative of the World Wide Fund for Nature (WWF), the United Nations Educational, Scientific, and Cultural Organisation (UNESCO) and the Royal Botanic Gardens, Kew, UK. Partial funding has been provided by the Darwin Initiative for the Survival of Species (Department of the Environment, UK).

Titles in this series

Ethnobotany: A methods manual
Gary J. Martin

Plant Invaders: The threat to natural ecosystems
Quentin Cronk and Janice Fuller

People and Wild Plant Use
Anthony B. Cunningham

Botanical Surveys for Conservation and Land Management
Peggy Stern and Peter Ashton

Botanical Databases for Conservation and Development

The cover illustration shows German Cayti, a Chimane indigenous person, and Quico Vaca, a park ranger, participating in an ethnobotany training workshop in the Beni Biological Station in Bolivia. They are preparing to dig roots of the palm *Scheela princeps* (used as a remedy for intestinal parasites) and demonstrating how to dry them for phytochemical analysis. Photo: G. J. Martin.

ETHNOBOTANY

'PEOPLE AND PLANTS' CONSERVATION MANUALS

Manual Series Editor:
Martin Walters

Manual Series Originator:
Alan Hamilton
Plants Conservation Officer
WWF International

'People and Plants' is a joint initiative of the World Wide Fund for Nature (WWF), the United Nations Educational, Scientific, and Cultural Organisation (UNESCO) and the Royal Botanic Gardens, Kew, UK. Partial funding has been provided by the Darwin Initiative for the Survival of Species (Department of the Environment, UK).

Titles in this series

Ethnobotany: A methods manual
Gary J. Martin

Plant Invaders: The threat to natural ecosystems
Quentin Cronk and Janice Fuller

People and Wild Plant Use
Anthony B. Cunningham

Botanical Surveys for Conservation and Land Management
Peggy Stern and Peter Ashton

Botanical Databases for Conservation and Development

The cover illustration shows German Cayti, a Chimane indigenous person, and Quico Vaca, a park ranger, participating in an ethnobotany training workshop in the Beni Biological Station in Bolivia. They are preparing to dig roots of the palm *Scheela princeps* (used as a remedy for intestinal parasites) and demonstrating how to dry them for phytochemical analysis. Photo: G. J. Martin.

ETHNOBOTANY

A methods manual

GARY J. MARTIN

WWF International
(World Wide Fund for Nature)

UNESCO
(United Nations Educational, Scientific, and Cultural Organisation)

Royal Botanic Gardens, Kew, UK

CHAPMAN & HALL

London · Glasgow · Weinheim · New York · Tokyo · Melbourne · Madras

Published by Chapman & Hall, 2–6 Boundary Row, London SE1 8HN, UK

Chapman & Hall, 2–6 Boundary Row, London SE1 8HN, UK

Blackie Academic & Professional, Wester Cleddens Road, Bishopbriggs, Glasgow G64 2NZ, UK

Chapman & Hall GmbH, Pappelallee 3, 69469 Weinheim, Germany

Chapman & Hall USA, One Penn Plaza, 41st Floor, New York NY 10119, USA

Chapman & Hall Japan, ITP-Japan, Kyowa Building, 3F, 2-2-1 Hirakawacho, Chiyoda-ku, Tokyo 102, Japan

Chapman & Hall Australia, Thomas Nelson Australia, 102 Dodds Street, South Melbourne, Victoria 3205, Australia

Chapman & Hall India, R. Seshadri, 32 Second Main Road, CIT East, Madras 600 035, India

First edition 1995

Typeset in Goudy 10/13 by ROM-Data Corporation Ltd., Falmouth, Cornwall

Printed in Great Britain at the University Press, Cambridge

ISBN 0 412 48370 X

A catalogue record for this book is available from the British Library

Library of Congress Catalog Card Number: 94–72447

∞ Printed on permanent acid-free text paper, manufactured in accordance with ANSI/NISO Z39.48-1992 and ANSI/NISO Z39.48-1984 (Permanence of Paper).

WWF

The World Wide Fund for Nature (WWF), founded in 1961, is the world's largest private nature conservation organization. It consists of 29 national organizations and associates, and works in more than 100 countries. The coordinating headquarters are in Gland, Switzerland. The WWF mission is to conserve biodiversity, to ensure that the use of renewable natural resources is sustainable and to promote actions to reduce pollution and wasteful consumption.

UNESCO

The United Nations Educational, Scientific, and Cultural Organisation is the only UN agency with a mandate spanning the fields of science (including social sciences), education, culture and communication. UNESCO has over 40 years of experience in testing interdisciplinary approaches to solving environment and development problems, in programs such as that on Man and the Biosphere (MAB). An international network of biosphere reserves provide sites for conservation of biological diversity, long-term ecological research, and testing and demonstrating approaches to the sustainable use of natural resources.

The Royal Botanic Gardens, Kew

The Royal Botanic Gardens, Kew has 150 professional staff and associated researchers and works with partners in over 42 countries. Research focuses on taxonomy, preparation of floras, economic botany, plant biochemistry and many other specialized fields. The Royal Botanic Gardens has one of the largest herbaria in the world and an excellent botanic library.

Darwin Initiative for the Survival of Species

At the Earth Summit in June 1992, the Prime Minister of the United Kingdom announced the Darwin Initiative as a demonstration of the UK's commitment to the aims of the Biodiversity Convention. The Initiative will build on Britain's scientific, educational and commercial strengths in the field of biodiversity to assist in the conservation of the world's biodiversity and natural habitats, particularly in those countries rich in biodiversity but poor in resources.

International Society for Ethnobiology

The ISE is the largest professional organization dedicated to sustainable use of natural resources and community development, and has endorsed the compilation of this manual.

DISCLAIMER

Contents

The 'People and Plants' Initiative

This manual is one of a series, forming a contribution to the People and Plants Initiative of the World Wide Fund for Nature (WWF), the United Nations Educational, Scientific, and Cultural Organisation (UNESCO) and the Royal Botanic Gardens, Kew (UK). The Initiative has received financial support from the Darwin Initiative for the Survival of Species (Department of the Environment, UK) and the Tropical Forestry Program (US Department of Agriculture). The main objective of the People and Plants Initiative is to build up the capacity for work with local communities on botanical aspects of conservation, especially in countries with tropical forests. The principal intention in the manual series is to provide information which will assist botanists and others to undertake such practical conservation work. Other components of the People and Plants Initiative include demonstration projects in Malaysia, Madagascar, Uganda, Mexico, Brazil and Bolivia, as well as support for workshops and publication of working papers.

This manual on ethnobotany is central to the People and Plants Initiative. It contains descriptions of methods useful for working with local communities to learn about their knowledge and uses of the plant world, for example to determine which social groups within a village use which species of plants in what quantities and for what purposes. Ethnobotanical studies can help identify conservation issues, such as cases where rates of harvest of plants exceed rates of regrowth. The fact that ethnobotany is, or should be, a collaborative venture between people in local communities, including various specialists, and scientists means that a start can be made to explore solutions to conservation and development issues, even as information on plant uses is being collected.

Many people, especially in poorer countries, rely on wild-collected plants for food, construction materials, fuelwood, medicine and many other purposes. Today, there is often a decrease in the availability of wild plant resources, related to increased human populations and the effects of competition with other forms of land use. Ethnobotanical surveys can help local communities define their needs for plant resources more clearly, thus assisting them to state their cases for continuing access to certain areas of land or for provision of alternatives to wild gathering, should this be necessary.

Traditionally, local communities worldwide are extremely knowledgeable about local plant and other natural resources, on which they are so immediately and intimately dependent. Unfortunately, much of this wealth of knowledge is today becoming lost as traditional cultures become eroded. Ethnobotanists can play very useful roles in rescuing disappearing knowledge and returning it to local communities. In this way local ethnobotanical knowledge can be conserved as part of living cultural-ecological systems, helping to maintain a sense of pride in local cultural knowledge and practices, and reinforcing links between communities and the environment, so essential for conservation.

Alan Hamilton
Plants Conservation Officer
WWF International

International panel of advisors

Janis Alcorn
Biodiversity Support Program
United States of America

Brent Berlin
Department of Anthropology
University of California at Berkeley
United States of America

Javier Caballero
Botanical Garden, Institute of Biology
National Autonomous University of Mexico
Mexico

Elaine Elizabetsky
Department of Pharmacology, Institute of Bioscience
Federal University of Rio Grande do Sur
Brazil

Richard Ford
Department of Anthropology
University of Michigan
United States of America

Timothy Johns
Centre for Nutrition and the Environment of Indigenous Peoples
McGill University
Canada

Virginia Nazarea-Sandoval
Department of Anthropology
University of California at Santa Cruz
United States of America

Darrell Posey
St. Anthony's College
University of Oxford
United Kingdom

Pei Sheng-ji
Kunming Institute of Botany
People's Republic of China

Ghillean Prance
Royal Botanic Gardens, Kew
United Kingdom

Internal panel of advisors

Alejandro de Avila
Field Representative, Mexican Program
WWF – US

Miguel Clüsener Godt
Man and the Biosphere Program
UNESCO

Phillip Cribb
Acting Head, Economic Botany and Conservation
Royal Botanic Gardens, Kew

Anthony Cunningham
Coordinator, People and Plants Initiative

Malcolm Hadley
Man and the Biosphere Program
UNESCO

Alan Hamilton
Plants Conservation Officer
WWF – International

Michel Pimbert
Head, Biodiversity Department
WWF – International

Nat Quansah
WWF Aires Protegées
Madagascar

Charles Stirton
Deputy Director
Royal Botanic Gardens, Kew

Foreword

Ethnoecology has blossomed in recent years into an important science because of the realization that the vast body of knowledge contained in both indigenous and folk cultures is being rapidly lost as natural ecosystems and cultures are being destroyed by the encroachment of development.

Ethnobotany and ethnozoology both began largely with direct observations about the ways in which people used plants and animals and consisted mainly of the compilation of lists. Recently, these subjects have adopted a much more scientific and quantitative methodology and have studied the ways in which people manage their environment and, as a consequence, have used a much more ecological approach. This manual of ethnobotanical methodology will become an essential tool for all ethnobiologists and ethnoecologists. It fills a significant gap in the literature and I only wish it had been available some years previously so that I could have given it to many of my students. I shall certainly recommend it to any future students who are interested in ethnoecology.

I particularly like the sympathetic approach to local peoples which pervades this book. It is one which encourages the ethnobotanical work by both the local people themselves and by academically trained researchers. A study of this book will avoid many of the arrogant approaches of the past and encourage a fair deal for any group which is being studied. This manual promotes both the involvement of local people and the return to them of knowledge which has been studied by outsiders.

Ethnoecology is by its very nature an interdisciplinary subject, demanding a holistic approach which integrates techniques from biology, anthropology, nomenclature, ethnology and many other fields. The author is well qualified to write a manual on such a broad field because he was trained in both botany and anthropology and consequently blends his two disciplines together in a well balanced approach to the subject. The use of this manual will certainly help to raise the standard of future work in ethnobotany.

Ghillean T Prance
Director
Royal Botanic Gardens, Kew

Preface

How did this manual come about?

I became interested in writing a book on ethnobotanical methodology while studying indigenous classification of plants in the Sierra Norte of Oaxaca, Mexico. When writing up the results of this research for my doctoral thesis, I met Alan Hamilton, Plants Conservation Officer of the World Wide Fund for Nature (WWF International). He considered ethnobotany to be an essential part of his program and together we sketched out a proposal to include this book in a series of WWF manuals on plant conservation.

After the manual was underway, WWF joined forces with the United Nations Educational, Scientific, and Cultural Organization (UNESCO) and the Royal Botanic Gardens, Kew to launch 'People and Plants', an international initiative that provides support and training for local ethnobotanists. Among other activities, the partner institutions in this initiative are sponsoring a series of training seminars that bring together local people, researchers and students to discuss how to carry out and apply ethnobotanical studies.

As part of the People and Plants initiative, I am continuing my work in Mexico and am participating in ethnobotanical projects in Kinabalu Park in Sabah – a Malaysian state on the island of Borneo – and in the Beni Biosphere Reserve, which is located in the Bolivian Amazon. These field experiences, as well as the contacts I have made with colleagues during seminars and visits in other countries, have much enriched this manual.

The concepts and techniques I present are drawn in small part from my own fieldwork and in large part from the research and publications of other ethnobotanists. From among these talented colleagues, I chose a group of international experts to advise me on the project. The panel's members, who are listed at the front of the manual, reflect the diverse origins and disciplines that characterize ethnobotany. In addition, several colleagues from WWF, UNESCO and Kew who participate in People and Plants – also listed at the front of the manual – reviewed the manuscript and contributed their valuable perspectives on the link between ethnobotany, biodiversity conservation and community development. Various colleagues have provided comments on specific chapters, including Michael

Balick, John Beaman, Chayan Picheansoonthon, Eugene Hunn, Oliver Phillips and Lionel Robineau. Participants in a graduate seminar on ethnobiology, organized by Brent Berlin at the University of California at Berkeley, reviewed the entire manuscript and provided many useful comments.

Malcolm Hadley and Alison Semple of the UNESCO Man and the Biosphere Program provided encouragement and technical support throughout the period of writing. Martin Walters, editor of the WWF plant conservation manual series, worked closely with me on the text, ensuring its readability. He also arranged for all of the illustrations, skilfully executed by Oxford Illustrators and Janet Simmonett. Although I drew upon the expertise of many people in writing the manual, I assume responsibility for its final form.

An autobiographical note and some acknowledgements

Because this manual reflects my own cultural and scientific biases, readers should know how I came to study ethnobotany and who supported me in this endeavor.

At first, I didn't wander far from home in my botanical explorations. I merely crossed the street in East Lansing, my home town, to study botany at Michigan State University. John Beaman, the curator of the University's Beal-Darlington Herbarium, invited me to collect plants with him in Mexico and Guatemala, an experience which converted me to tropical botany.

After finishing my bachelor's degree, I returned to Mexico to pursue a lingering dream left from my first trip – to discover which plants were being used for food and medicine or traded in rural marketplaces by indigenous communities in Oaxaca. After a year in Mexico, I opted to pursue a doctorate in anthropology at the University of California at Berkeley, where I focused on ethnobiology.

When it came time to pursue my dissertation research, I returned to Oaxaca to work among the Chinantec, Mixe and Zapotec Indians of the Sierra Norte. My theoretical approach was guided by Brent Berlin, a professor of anthropology at Berkeley who has written extensively on the folk classification of plants and animals. I owe my methodological orientation to Eugene Hammel, a Berkeley demographer who taught me how to use computers, statistics and common sense in designing my research.

In Mexico, I had the opportunity to work with Stefano Varese, a Peruvian anthropologist who was coordinating the Oaxaca regional office of *Culturas Populares*, a program of Mexico's *Secretaría de Educación Pública*, which brought together cultural promoters from Chinantec, Mixe and Zapotec communities. While interacting with Stefano and his co-workers, I came to appreciate the complexities of applying basic ethnobotanical research to community development.

Towards the end of my stay in Oaxaca, several colleagues and I began to form a non-profit group that eventually became known as SERBO: *Sociedad para el Estudio de los Recursos Bióticos de Oaxaca* (Society for the Study of the Biotic Resources of Oaxaca). SERBO's focus on integrating research with efforts towards biodiversity conservation and community development has been a constant source of inspiration for this manual.

The writing of this manual was made possible through a grant from WWF International, with additional support from UNESCO. The field projects in Mexico, Malaysia and Bolivia in which I participate have been supported by various foundations. The ethnobotanical survey of the Sierra Norte of Oaxaca was funded by grants from the Garden Club of America, National Science Foundation (Doctoral Dissertation Improvement Award), Wenner-Gren Foundation for Anthropological Research and World Wildlife Fund (US). I was supported during my tenure in Mexico by Fulbright-Hays and National Science Foundation pre-doctoral training fellowships and by a scholarship from the Inter-American Foundation. Additional support was provided by the University of California at Berkeley, Missouri Botanical Garden and New York Botanical Garden. The *Projek Ethnobotani Kinabalu* in Malaysia was initiated through grants from WWF International and UNESCO, and is continuing with support from the MacArthur Foundation. My participation in the *Proyecto Etnoecológico del Beni* in the Beni Biological Station of Bolivia began through a Fulbright award from the Council for the International Exchange of Scholars of the United States. Additional support has been provided by UNESCO.

Without the hospitality and collaboration of many local people in communities in Latin America and Asia, I would never have realized my goal of becoming an ethnobotanist. I would like to thank in particular the people of Santiago Comaltepec and Totontepec who helped me during my first intensive study.

A dedication

A. Barrera, one of the first promoters of ethnobotany in Mexico, wrote:

> The best ethnobotanist would be a member of an ethnic minority who, trained in both botany and anthropology, would study . . . the traditional knowledge, cultural significance, and the management and uses of the flora. And it would be even better – for him and his people – if his study could result in economic and cultural benefits for his own community.

This book is dedicated to those aspiring ethnobotanists who set out to record and apply ecological knowledge among their own people.

The manual is also dedicated to my family, who have endured with patience my long hours at the computer and my frequent trips abroad. I hope that they will increasingly share in the fruits of my work.

Paris, France
February 1994

Introduction

Ethnoecology – the broader discipline?

Ethno- is a popular prefix these days, because it is a short way of saying 'that's the way **other people** look at the world'. When used before the name of an academic discipline such as botany or pharmacology, it implies that researchers are exploring local people's perception of cultural and scientific knowledge. Leaf through the index of an anthropological journal and you will find articles that range from ethnoastronomy (local perception of stars, planets and other celestial bodies) to ethnozoology (local knowledge and use of animals). The term **ethnoecology** is increasingly used to encompass all studies which describe local people's interaction with the natural environment, including subdisciplines such as ethnobiology, ethnobotany, ethnoentomology and ethnozoology. **Ethnobotany is that part of ethnoecology which concerns plants**.

How can ethnoecology be described to someone who has never heard the term? The basic definition is the one implied above – the study of how people interact with all aspects of the natural environment, including plants and animals, landforms, forest types and soils, among many other things [43,73,141,149,159]. Given this diversity of subjects, ethnoecology is a multidisciplinary endeavor which attracts a broad community of people who contribute their own special knowledge and skills [20,140]. At times, several researchers work together as a team in planning and carrying out a field project. Botanists collect and identify plant specimens, linguists study local names, anthropologists record ecological knowledge, pharmacognosists analyze medicinal properties of plants, ecologists describe the local classification of vegetation types, zoologists monitor animal populations, resource economists estimate the value of forest products – and the roster goes on.

This teamwork is the exception rather than the rule. Ethnoecologists must often labor alone, trespassing the boundaries of many academic disciplines in their quest to attain a holistic vision of local ecological knowledge. Because of some romanticized accounts, there persists a popular image of ethnoecologists as loners who venture into unexplored virgin forests to contact isolated groups of indigenous people and to make lists of the medicinal and hallucinogenic plants of which only shamans and witch doctors know the secrets. Some observers conclude that ethnoecology is an old-fashioned business of creating catalogues of useful plants and animals, something more associated with early explorers, missionaries and natural historians than with modern scientists.

There are many elements of these myths that must be dispelled. Although they work in primary forests, ethnoecologists are also interested in a broad range of vegetation types which have been altered by people, ranging from home gardens to mature secondary forests, where the majority of useful plants are found.

Who are these local people that ethnoecologists study? Much research focuses on the ecological knowledge of the world's indigenous people, but some ethnobotanists work with traditional agriculturalists who do not consider themselves 'indigenous'. Those who do define themselves as indigenous have varied lifestyles. While it is true that some have had little contact with outsiders, the majority are integrated into the economy and politics of the country in which they live. Although most ethnoecology is done in rural areas, some studies – such as the description of plants or animals sold in urban marketplaces – are carried out in cities.

Ethnoecology is no longer concerned with mere list-making. Systematic research into local ecological knowledge allows us to address theoretical questions about the relationship between humans and their environment and to contribute to rural development and conservation projects. As part of these theoretical and practical approaches, ethnoecologists continue to catalogue information on the local classification and use of plants and animals. Any impression that these ethnobiological inventories are old-fashioned will be quickly dispelled by listening to local people and researchers who tell of the urgency of recording ecological knowledge and collecting biological organisms before they disappear forever.

As a discipline which integrates many diverse academic fields, ethnoecology is having an impact on the way that basic and applied research is carried out. After decades in which the natural sciences and even the social sciences have become increasingly reductionist, ethnoecology promises to give a holistic view of our knowledge of the environment. This basic goal of natural history, nearly abandoned in the quest for specialized knowledge in fields such as particle physics and molecular genetics, is gaining new importance as scientists seek to understand the ecological wisdom of local people.

Ethnobotany – the scope of this manual

Because ethnoecology is too broad a subject to be condensed into a single book, this manual focuses on ethnobotany and in particular on methods of making inventories of useful plants. Even this is a very broad subject and ethnobotanists need to have some understanding of a broad range of academic disciplines. Ethnobotanists often have to work without the support of colleagues in other subject areas – and there can be advantages in doing so from the point of establishing close relationships with local communities. However, in ma stances there are advantages in the team approach, involving specialists in various disciplines such as plant taxonomy, anthropology, linguistics, economic botany and others, working together to achieve more detailed and reliable results.

This manual describes the basic concepts, skills and techniques that guide collection of quality data in the field. The style and content are designed for field workers – including park rangers, university students, cultural promoters and nature guides – who are beginning their first research project. I hope the manual will also be useful to seasoned ethnobotanists and colleagues of related disciplines.

The manual is to be used in designing projects which yield not only accurate information, but also practical results that can be applied to community development and biological conservation. It is intended for those who need to gather ethnobotanical data in a few weeks or months as well as those who can dedicate many years of research to a single area. Although focused on fieldwork, the manual offers advice on ways of following up field studies by consulting herbaria, libraries and museums as well as working with specialized colleagues in research laboratories.

In order to make the manual accessible to people from various educational backgrounds, I have left the text free of conventions of scientific writing such as footnotes. I mention by name the many researchers who have contributed to the ideas presented in the manual and in the bibliography I list some of the books and articles they have written. All technical terms are **highlighted** and defined when first used. Terms that are transcribed from indigenous languages are both highlighted and italicized, such as **ojts**, the Mixe word for 'herb'. Words from languages such as Spanish and French are simply *italicized*.

The first chapter gives a basic description of data gathering and hypothesis-testing. Chapters 2 through 7 form the core of the manual, exploring the contribution to ethnobotany from the diverse fields of botany, ethnopharmacology, anthropology, ecology, economics and linguistics. Chapter 8 explores the link between ethnobotany, biodiversity conservation and community development.

Throughout the manual, I often turn to the case studies I know best – the ethnobotanical inventories I am carrying out with Chinantec and Mixe-speaking people in the Sierra Norte of Oaxaca and with Dusun people who live around Kinabalu Park in Sabah, Malaysia. The Sierra Norte, approximately 300 kilometers long and 80 kilometers wide, ranges from 320 meters to nearly 3400 meters above sea level. The region comprises 3 *distritos* and 59 *municipios* that are found between 16°45′ N and 18°10′ N latitude and 96°06′ W and 98°30′ W longitude, south of the Tropic of Cancer. Santiago Comaltepec, a highland Chinantec municipality in the Sierra Norte, includes some 2000 residents who inhabit nearly 200 square kilometers of communal lands and private property. Totontepec is a highland Mixe municipality which includes some 320 km^2 of territory and over 5300 inhabitants.

Mount Kinabalu – in Sabah, Malaysia – is the highest mountain between the Himalayas and New Guinea. Rising to 4101 meters in northern Borneo, it is the centerpiece of a protected mountainous zone that measures approximately 700 km^2. Around the perimeter of the park, there are many communities of Dusun-speaking people.

By referring to these localities in the various chapters, I hope to demonstrate the interrelationship of the many different aspects of local people's knowledge of the environment. In addition, I take case studies from work carried out by many colleagues throughout the world. By indicating the diversity of approaches and cultural contexts of our research, I hope to show that ethnobotanical studies can be carried out by a variety of people in diverse countries and ecosystems.

Due to limitations of space, the original manuscript was trimmed of much background and supplementary material. Many of these sections have been included in a report in the MAB Digest series, provisionally entitled *Comparing Scientific and Traditional Botanical Knowledge*, which can be ordered from the People and Plants Initiative, Division of Ecological Sciences, Man and the Biosphere Program, UNESCO, 7, place de Fontenoy, 75352 Paris Cedex 07 SP, France. It is hoped that the Initiative will publish a number of smaller booklets to supplement this manual, including a sourcebook on how to find bibliography and financial support for projects, as well as guides to topics which could not be covered in detail in this text, such as phytochemical screening in the field and comparison of traditional and scientific diagnosis of illness.

A note about terminology

Ethnobotanical fieldwork was originally conceived as an art and skill practised by outsiders who travelled to distant lands to document customs and beliefs. At present, many people have adopted ethnobotanical methods to carry out studies of their own communities. Although this manual is intended both for outside researchers and for local people who are studying their cultural beliefs, it is difficult to write a description of ethnobotanical methods without sometimes referring to local people as the objects of the study and researchers as the people who carry out the investigation.

I employ the term **local people** for residents of the region under study who have gained their ecological knowledge from empirical observation of nature and from communication with other people in their culture. To be more precise in specific examples, I sometimes use the terms indigenous people, traditional agriculturalists or subsistence farmers as synonyms.

I use **researchers** for people, usually trained at a university, who document this traditional knowledge in collaboration with local people. In the context of certain case studies, researchers are called by the name which corresponds to their chosen profession – botanists, anthropologists, linguists, or simply scientists.

Much of the specialized vocabulary that ethnobotanists employ is marked by old prejudices, which makes it difficult to describe the participants in our joint venture. Calling one group 'scientific' implies that the empirical knowledge of the others lacks rigor, whereas labelling some people 'traditional' may promote the mistaken notion that their partners are more modern.

Throughout the text, I use the terms **traditional knowledge** or **folk knowledge** to refer to what local people know about the natural environment, whereas I consider **scientific knowledge** as information derived from research. These terms are widely understood and accepted by ethnobotanists, who understand that traditional knowledge has sometimes been acquired recently, some scientific knowledge is derived from Indo-European traditional beliefs and folk knowledge demonstrates much scientific rigor [8,68,158].

In creating a dichotomy between scientific and folk knowledge, we are apt to evoke both old stereotypes and new controversies. Some people think that science provides the correct model of the natural world and should be adopted by everybody. For this reason, they give little credit to self-taught naturalists who lack a formal education. Others feel that the right to cultural self-determination is pre-eminent and that science can only corrupt the purity of indigenous knowledge.

Ethnoecologists do not seek to judge systems of knowledge, declaring one superior to another. Their research has revealed both the wealth of detailed information contained in folk systems of natural science and the utility of using scientific classification as a looking glass through which various indigenous systems can be observed and compared. By employing the techniques described in this manual, we can bring our understanding and appreciation of traditional knowledge into sharper focus.

1

Data collection and hypothesis testing

Figure 1.1 German Cayti, from the Chimane indigenous community of Puerto Mendez, Bolivia, showing scarring on the bark of *Hura crepitans* (Euphorbiaceae), which is used as a fish poison. Local people can provide a vast amount of data on local plant resources.

1.1 Choosing an approach

When designing an ethnobotanical project, it is important to define what you wish to achieve and then to select the approach which best suits your interests, budget and schedule. It is easy to be overly enthusiastic about what can be accomplished in a short field season. Once the project begins, you discover the complexity of local ecological knowledge and the diversity of the flora and fauna. You experience unforseen delays caused by the weather, equipment failure and other events beyond your control.

Many of the methods used in ethnobotanical studies are time-consuming and moderately costly, making it impractical to apply all in a single period of fieldwork. For this reason many researchers divide their time between visits to the field with stays back home analyzing the data and writing up the results. Once you complete the process of adapting a technique for your field site, collecting the data and analyzing the results, you will be better able to choose a complementary approach in the future. For example, if you record the uses and names of medicinal plants in a community, you can identify the specimens and search ethnopharmacology databases and literature to evaluate which merit further study before returning to the field to gather samples for chemical analysis.

It may be costly to make several trips to the field, but remember that the most satisfactory projects – from a personal viewpoint as well as from the perspective of the community and your colleagues – are those which span several seasons and continue for a number of years.

In practice, the time and resources that ethnobotanists can dedicate to projects vary widely. Let's take the example of several case studies discussed in this manual:

- K.C. Malhotra, M. Poffenberger, A. Bhattacharya and D. Dev spent two days making a rapid inventory of non-wood forest products and an assessment of forest regeneration in a village in West Midnapore District in Southwest Bengal. This formed part of a long-term study of the impact of Forest Protection Committees on forest regeneration in the region [1].

- O. Phillips spent a total of 12 months over a period of five years in the Tambopata Reserve in southern Peru to document Mestizo use of plant resources [2–4].

- R. O. Guerrero and I. Robledo, combining work in the field and the laboratory, analyzed the biological activity of plants from the Caribbean National Forest over a period of 30 months [5].

- B. Berlin worked with various colleagues in Chiapas, Mexico, for several years in the early 1960s to document the botanical classification of the Tzeltal, a group of Maya-speaking agriculturalists. Berlin returned to the region in the late 1980s with other colleagues to focus on the medical ethnobiology of the highland Maya in a multi-year project [6, 7].

2

1.2 Six disciplines which contribute to an ethnobotanical study

Because it is often said that ethnobotany is a multidisciplinary endeavor, it should be easy to enumerate the fields of study that contribute to analyzing how humans interact with the plant world. What are they? This is a question often posed by people who are beginning their first ethnobotanical study. Yet even experienced researchers fumble about for the recipe: botany of course and some linguistics, a background in anthropology helps and you had better know some chemistry and economics.

The response depends in part on the kind of project that is planned. There are four major interrelated endeavors in ethnobotany: (1) basic documentation of traditional botanical knowledge; (2) quantitative evaluation of the use and management of botanical resources; (3) experimental assessment of the benefits derived from plants, both for subsistence and for commercial ends; and (4) applied projects that seek to maximize the value that local people attain from their ecological knowledge and resources. Walter Lewis, an ethnobotanist at Washington University, has suggested that the first three elements be referred to as basic, quantitative and experimental ethnobotany, respectively [8].

The focus in this manual is on six fields of study: botany, ethnopharmacology, anthropology, ecology, economics and linguistics. In any long-term project, techniques borrowed from these fields can be combined to carry out a systematic survey of the traditional botanical knowledge in a single community or region.

1.3 Rapid ethnobotanical appraisal

Although many researchers prefer long-term projects, they are sometimes called upon to make a rapid ethnobotanical study – gathering data on minor forest products for an environmental impact statement, making a preliminary list of biological resources at sites that have been set aside as protected areas or simply conducting an initial ethnobotanical inventory in several communities in order to decide where it would be most interesting to carry out long-term research.

We can find many faults with studies that only last a few days. They do not allow a deep working relationship to develop between an ethnobotanist and the community. Careful documentation of the cultural and biological aspects of local knowledge is not possible, because there is little time to make voucher collections, transcribe local names or talk with a range of informants. Above all, short visits do not permit local people to learn rigorous ethnobotanical methods that would allow them to manage more effectively resources in their own communities. Yet the urgency of encountering solutions to community and conservation problems sometimes requires that we make a quick assessment of ecological knowledge and resource use, while rapidly teaching local people some of the basic techniques we employ.

As a response to this dilemma, international development workers have improvised various methods of making a fast low-cost assessment of the use of forest resources and many other aspects of community development. Techniques adopted from various disciplines have been combined to form a collaborative approach called Participatory Rural Appraisal (PRA). Although originally developed to guide and evaluate development initiatives, PRA is readily applicable to ethnobotanical studies, as can be seen from the example given in Box 1.1.

Box 1.1 Rapid appraisal of non-wood forest products in India

Over a two-day period in 1991, a multidisciplinary team carried out a rapid appraisal of forest regeneration and harvesting of non-wood forest products with the members of a community in West Midnapore District in Southwest Bengal. As a first step, the participants selected several 100 m^2 plots that showed various degrees of protection from deforestation. They made an inventory of the trees in each plot, determining the species and recording the size of each individual.

In a working paper, K.C. Malhotra, M. Poffenberger, A. Bhattacharya and D. Dev [1] describe the ethnobotanical component of this experience:

... tribal and some non-tribal people who have strong ties to the forest can identify hundreds of productive species and how they are used as sources of foods, medicines, fiber and construction materials, gums, dyes, tannins, etc. Using secondary data and local resource persons this information was documented in the case study sites. Listings were made for all products used for home consumption or sale. Data were collected regarding harvesting seasons and volumes. An attempt was also made to determine the parts of the plant utilized.

During this rapid appraisal, many useful plant species were identified. For example, the researchers recorded the names of 29 minor forest products available in a forest that had been protected for three years. The Latin names of 15 of these were noted and information on the market price was recorded for six of them.

In their report, the coordinators of this participatory exercise also point out the limitations of their assessment. They were not able to document all of the important biological resources, neglecting for example the mushrooms that are available only during the monsoon and early post-monsoon periods. They did not enquire in a systematic way about the volume of plant materials harvested or what prices they brought in local markets. No ethnobotanical voucher collections were made and the scientific names of nearly half of the plants went undetermined.

Judging the successes and deficiencies of the experience, they made suggestions on how future research should proceed. Among other recommendations, they proposed that the exercise be extended to include visits during the various seasons of the year and that an intensive study be made of the marketing of non-wood resources, including collection of data on volume flows and the possible depletion of certain plants.

Although PRA borrows many of its tools from traditional disciplines such as rural sociology, anthropology, ecology and economics, there are important differences that distinguish this rapid approach from academic research. Local people are full participants in the study rather than being merely the objects of the investigation. They take part in the design of the study, data collection, analysis of the findings and discussions of how the results can be applied for the benefit of the community. The outsiders in the research team come from a variety of academic backgrounds, ensuring a multidisciplinary perspective. The relationship between all participants – locals and outsiders – is egalitarian, avoiding the hierarchical or top-down approach common to much research.

The techniques can be carried out in a short time and do not require expensive tools because participants are seeking a sketch of local conditions rather than an in-depth study. A small group of local people is selected to be interviewed in a semi-structured way. A wide range of topics may be covered in a preliminary way, allowing a comprehensive view of how the community works as a whole. Measurements are qualitative rather than quantitative and few statistical tools are used in the interpretation of the results. The emphasis is on highly visual techniques that community members carry out amongst themselves, often in collaboration with outside researchers – sketching maps to show local classification of ecological zones, creating pie charts that represent the amount of time that people dedicate to various productive activities or drawing calendars which show seasonal fluctuations in climate, to give just a few examples. The analysis of the data is carried out in the community, which allows participants to modify their methods on the spot and to fill in any data which are missing after initial fieldwork. Often the participants pass through successive rounds of data gathering and analysis, which allows them to refine their techniques during the course of the exercise.

PRA is a cost-effective approach because much can be accomplished in a few days, including writing up of the final results and recommendations. This stands in contrast to many traditional methods of field research which require months of data analysis and writing before final conclusions can be reached.

These various aspects of PRA guarantee a good amount of flexibility, allowing the approach to be adapted to the very diverse cultural and ecological conditions under which ethnobotanists work. Throughout the manual, research tools from PRA are suggested as one way of documenting how people interact with their environment, especially as a complement to more rigorous methods of data collection.

Here are a few general points to keep in mind when planning a rapid ethnobotanical assessment, drawn from the experience of colleagues who have carried out participatory appraisals in different countries and cultural contexts [9, 10]:

- **Prepare yourself before fieldwork.** Obtain secondary information – maps, floras, faunas, vegetation analysis, census statistics, reports on forest use – to gain a preliminary idea of the land, the people and the conservation issues in the region. Consult the maps to select the specific sites and villages that you will visit. Whenever possible, have local people participate in the collection of these secondary sources of information, whether in their community or in nearby governmental offices or universities.

- **Form a multidisciplinary team.** Make prior contact with a linguist familiar with the local language, a botanist who knows the flora, an anthropologist who has studied local people's classification and management of the environment and other researchers who have worked in the region. If these people or some of their colleagues cannot accompany you in the field, see if they can give tips on what plants to collect, how to transcribe the language and what native categories exist. Determine if they can review the results of the appraisal after the fieldwork is finished or participate in a subsequent stage of the work. These preliminary consultations should be made by a small committee that includes local people, when possible.

- **Ensure community participation.** As a first step, seek the full cooperation and permission of local authorities before starting fieldwork. Ask them to recommend several local people, including specialist resource users, who can work with the assessment team. A general presentation can be made in the community to explain the goals of the appraisal, particularly if many families will be participating.

- **Be selective in your choice of techniques.** Concentrate on the methods that will yield the information you need for the appraisal, rather than exploring a broad range of techniques. Collect the minimum amount of data that will allow you to assess local patterns of resource use or ecological knowledge rather than trying to investigate a specific topic in great detail. Choose methods of analysis that can be understood by all participants and that do not require time-consuming calculations or expensive tools such as computers.

- **Do everything systematically.** The appraisal should be done systematically so that others who wish to conduct a more thorough study consult your results and add to them. This includes making a map of the sites that you visit, recording the names of all local people who participate in the exercise, identifying as fully as possible the biological species encountered and recording on paper or on tape the semi-structured interviews that are carried out. The conclusions of the appraisal, together with any drawings, charts or graphics created during the stay, should be presented in a final report that is written in a style accessible to a wide range of people, including local participants.

1.4 Planning a long-term project

After you carry out a rapid appraisal, you may decide to continue research for a few weeks, a season or several years during which you will be able to apply more rigorous research methods. A longer period in the field will allow you to work with local people to record ecological knowledge in a variety of social contexts, including community festivals, ritual occasions and seasonal farming activities. Regardless of the special techniques chosen and of the scope of the research, there are minimum standards of ethnobotanical documentation that should be considered in any long-term community project (explained in more detail in later chapters):

- Specimens of all species represented in the study must be collected, identified and deposited in an herbarium, zoological museum, seed bank or equivalent facility. The collections should be prepared in accordance with the guidelines supplied by collaborating biologists. All specimens should be accompanied by a label detailing the scientific name, locality, description, collector and number of the collection as well as other information.
- All local categories of plants should be identified and information collected on the distribution, use and management of the corresponding botanical species. This cultural information should be confirmed in discussions with numerous community members who represent the social diversity of the community where you are studying – rich and poor, young and old, men and women and so on. The age, educational background, occupation, language ability and other personal data should be recorded for each local person who participates in the project.
- All local plant names and other key terms must be accurately transcribed using a widely-accepted phonetic alphabet or local writing system. The names may be recorded on tape by several native speakers so that other researchers can review the accuracy of the transcriptions.
- Each plant or animal population sampled for analysis in a laboratory or research center should be documented by a voucher specimen. The material used for analysis must be prepared in a standard way in the field, appropriate for the tests that will be carried out.
- The local perception and classification of diverse aspects of the natural environment should be recorded, including concepts of vegetational communities, soil types, geographical landmarks, climate zones and seasons. The more you intend to study the management of natural areas, the more you must document the physical and biological aspects of the environment by taki

soil samples, measuring the species diversity of different ecological zones, describing the stages of ecological succession and so forth.

- The economic value of biological resources should be estimated, particularly if you plan to compare the economic advantages of different patterns of land use. You may record the price and availability of plant and animal products sold in rural marketplaces, assess the time people spend in harvesting the resources and estimate the cost of transporting the goods to where they can be sold.

1.5 Describing the field site

When you decide to carry out long-term research in an area, you should consult secondary sources such as ethnographies, maps and geographical accounts to describe the land, local people and conservation status of the region where you will be working. Table 1.1 lists some elements to be included in the description.

The geographical location of the research site should be stated, giving the country, state, province, municipality or other political geographical unit used locally. Citing the surface area in square kilometers (km^2) aids in comparison with other research sites. Reporting the longitudinal and latitudinal coordinates helps other researchers find the area on a topographical map. The history of the geological formation of the land should be described as well as the topography, soil types and geographical landmarks (watersheds, mountains, caves and so on). Delineate the major climatic zones, pointing out seasonal fluctuations in temperature and precipitation. Finally, characterize the floristic and faunistic regions and the major types of vegetation that cover the land, including the various stages of succession observable in the study site.

State the size of the human population and give information on its geographical and demographic distribution by providing answers to questions such as: What are the major settlements and where are they located?; How many inhabitants are men and how many are women?; How many people are there per household? When archaeological and ethnohistorical data permit, delineate the history of the people's arrival in the region, giving an estimate of the number of years they have been settled there. Record the languages they speak, including rates of monolingualism and bilingualism and levels of literacy and formal education. Describe their ethnicity, giving the approximate population of each ethnic group.

Characterize the degree of social stratification of the community and the relations between the various groups. Enumerate the productive activities of the local people and describe how the labor is divided between men and women, young and old and other social groups. List the important subsistence and commercial crops, estimating yields. Characterize the system of land tenure and how the territory is divided up into different categories of ownership. If people are migrating into or out of the region, give the origin or destination of the immigrants and estimate their number.

Table 1.1 Suggested categories for describing the land, people and conservation status of a research site

---

Land
Geographical location and map
Surface area in km^2
Longitude and latitude
Geological formation
Elevational range
Major geographical landmarks
Soil types
Climatic zones and seasons
Vegetation types and successional stages

People
Population size and distribution
Language(s) spoken
Ethnic affiliation(s)
History of settlement
Major social groups or classes
Productive activities
Subsistence and commercial crops
System of land tenure
Rates of migration

Conservation status
Size and status of protected areas
Transportation infrastructure
Natural and human-caused disasters
Colonization
Agriculture
Logging
Alternative land-use schemes
Nature tourism
Extractivism

After describing the land and the local inhabitants, characterize the conservation status of the region. Give the number and size of all areas that are officially protected, such as biosphere reserves, national or state parks or buffer zones. Note any traditional system of protected areas that are maintained by local people. Indicate current threats to both protected and unprotected sites. Describe natural and human-caused disturbances that affect broad areas in the region, including fires, floods, erosion and long-range sources of pollution such as fertilizer runoff, pesticides and acid-rain. Report the extent of transportation infrastructure – roads, airstrips – and how it affects the ease of access to natural areas. Characterize demographic trends and their impact on natural resources and vegetation: Is there large-scale colonization, driven by resettlement and immigration of people from another region?; Are these colonists destroying primary forest? Assess the ecological damage caused by subsistence and commercial agriculture, giving special attention to activities which result in large-scale clearing of the vegetation, such as cattle-raising, growing of export crops in plantations and shifting cultivation. Delimit the areas where logging is practised and characterize the species that are targeted. Calculate the amount of each that is harvested and determine who is processing and profiting from the timber (community-based sawmills, private pulp mills, international traders in valuable hardwoods and so on). Record other activities which threaten large tracts of land, such as mining and hydroelectric projects. When working in an area that is visited by tourists, assess their numbers and impact on the environment. If non-timber forest products are being extracted from the region, state who is doing the collecting (local people, scientific expeditions, hobbyists, commercial enterprises), the species affected and the approximate amount of material that is being removed. When possible, investigate if these are legal or clandestine activities.

1.6 Ethnobotanical data

1.6.1 What are ethnobotanical data?

When ethnobotanists speak of **data**, they are referring to the broad range of information they collect on how local people interact with the natural environment. Data are recorded in many different forms – collections of plants and animals, recorded interviews, laboratory analyses, photographs, market surveys and so on. A **data set** is a collection of information gathered in a systematic way – a list of all local firewoods elicited from several members of a community, the results of screening for alkaloids in selected medicinal herbs or a census of trees in a one hectare plot, to list just a few examples. Before setting out to compile a data set, we first define the **domain** or subject of interest. For example, we might delimit the domain by choosing to study all plants used for stomachaches, insects that are edible or categories that people use to describe local soils.

By collecting data sets in a systematic way using an explicit methodology, we can show colleagues how we arrived at our conclusions. Other researchers can then reanalyze our data or collect similar data in another area to see if they achieve the same results. In this way, we can see if our observations apply, for example, to a single community, to all traditional agriculturalists or to all people around the world.

1.6.2 Compiling a data set

It is never possible to compile a complete ethnobotanical data set. Imagine the difficulties. Every plant and animal in the community or region would have to be collected. You would need to accompany every man, woman and child into the field, asking them the names and uses of all organisms and observing how they prepare and use each species. You would be required to transcribe and analyze every folk name as well as all discussions, myths and legends that are in any way associated with what people know about the natural environment. All edible and medicinal plants would have to be analyzed in the laboratory. Animal populations would be monitored to assess the impact of hunting. Other ecological methods would be applied to assess the impact of plant resource harvesting. Such a process would be a burden for community members and would take more time, money and patience than ethnobotanists have at their disposal.

What shortcuts do we take to reduce our workload yet still achieve accurate results? First, we define carefully the scope of our study. Then we choose the techniques that will be employed and select a subsample of people who will participate in the project. Suppose we want to discover which plants are being cut for firewood in a certain village. We would need to collect only trees and shrubs, leaving the herbs, ferns, vines and other plants for another study. If we learn that only women and pre-adolescent children are responsible for fuel gathering, we could limit our questioning to a representative number of people selected from this segment of the community. We would transcribe only a small number of plant names. Our laboratory studies would be limited to discovering how much heat is emitted by each type of wood, if the wood burns green or must first be dried and other similar questions. This is already a time-consuming study, but much less demanding than an exhaustive ethnobotanical assessment.

Another important step in compiling a data set is to define how to categorize and measure local knowledge and management of the environment. **Categorization** is the way that something is divided up into a set of different classes, such as types of soils, lifeforms of plants and so on. In ethnobotanical studies, it is preferable to collect data by using **emic categories**, those drawn from the way people perceive things through their own eyes and classify objects in their own language. For instance, it is better to refer to the local names and symptoms of illnesses when interviewing people about medicinal plants than to employ a typology of health conditions drawn from Western medicine.

On occasion, you will also use **etic categories**, which are taken from the way the researcher perceives and classifies the world but do not necessarily form part of the classification systems of local people. For example, ethnobotanists commonly divide the use of plants into categories such as food, medicine, firewood and construction materials, even though these may not be distinct classes for the local people with whom they are working. If you must use etic categories, choose a simple set of classes that all participants in the study can grasp intuitively. Allow local people to modify the categories, adding new ones when they feel it is necessary. Above all, be careful not to force other peoples' knowledge into your own way of seeing the world without first trying to understand local perception and classification.

Units of measure allow us to quantify our observations, a concept familiar to

11

us from our everyday life. We use units such as years when speaking of a person's age, meters when we talk about the elevation of a mountain or kilograms when we weigh ourselves. The choice of units we select for a study always should be guided by common sense, custom and the relative need for precision in our results. There may be local systems of units that allow for greater accuracy of measurement in the field, simply because participants are accustomed to the tools and scales they use for measuring. These measurements can later be converted to an internationally recognized system of units to make the results comparable with other studies.

Before you begin measuring something, be clear on the degree of precision that you wish to attain and stick to the same units, tools and scales throughout the study. Fine measurements can always be lumped into grosser classes, but it is never possible to split broader measurements into finer divisions. For example, if you measure the diameter of a tree trunk to the nearest millimeter, you will have no problem dividing the trees into diameter size classes such as 10–15 cm, 15–20 cm and so on. However, if you use such broad classes to begin with, you will never be able to carry out calculations based on the precise diameter of the tree, such as calculating the average diameter size for a certain species.

1.6.3 Organizing the data

As your fieldwork progresses, you may have data of many different types, including information in collection notebooks, transcriptions or recordings of interviews, results from sorting tasks, plant identifications sent by collaborating botanists and laboratory analyses contributed by ethnopharmacologists. The next step is to organize this information in a way that will allow you to ask questions of the data and to compare the results of one part of the study with another.

Many researchers create a card file, a box of note cards on which they collate and cross-index information. For example, in a card file organized according to biological species, a separate note card is made for each different plant and animal and is then placed in alphabetical order. All information pertaining to the species is written on the card – the collection numbers of specimens, local names, uses and so on. This card file allows the researcher to see if the species are named and used in the same way by all local participants.

What if the ethnobotanist wants to know which local names correspond to a single biological species or which plants are used to treat skin diseases? Additional card files would have to be made, one ordered by indigenous name, the other by use. Ideally, each card file would be cross-referenced to the others. An example of this kind of card file is described in Box 1.2.

Box 1.2 A card file on Tzeltal ethnobotany

In the early 1960s, one of the first major inventories of folk botanical categories was initiated by an anthropologist, Brent Berlin, and two botanists,

Dennis Breedlove and Peter Raven. During several years of fieldwork in Chiapas, a state of southern Mexico, they collected over 10 000 plants and documented many hundreds of folk botanical categories of the Tzeltal, a Maya-speaking group of the highlands of Chiapas.

In their monograph, *Principles of Tzeltal Plant Classification*, Berlin and his colleagues [7] describe their use of a card file that brings together botanical, linguistic and ethnographic data:

> ... a series of Tzeltal collection files were begun early in the research. These files were drawn directly from our Tzeltal collection notebooks and arranged arbitrarily in alphabetical order. A collection file entry, made on 4 × 6-inch slips, would include the Tzeltal name and all collection numbers of that plant with their corresponding informant identification numbers... a duplicate file, made on regular 8.5 × 11-inch paper, was begun, which included the botanical identifications, as these became available.

Berlin and his colleagues give an example, depicted in Table 1.2, of one card from the file ordered by indigenous name. The card includes the Tzeltal name, the collection number, the number of the informant who identified the collection and the botanical determination.

Table 1.2 Example of a card from the Tzeltal Maya ethnobotany card file created by Brent Berlin and his colleagues during fieldwork in Chiapas, Mexico

Tzeltal name	Collection number	Informant's number	Botanical determination
'ac'am te'	204	8	*Rapanea myricoides*
	499	8, 53, 54, 55	*Rapanea myricoides*
	587	8, 56	*Rapanea myricoides*
	6876	2	*Myrica cerifera*
	6954	3	*Rapanea juergensenii*
	7032	1, 2A, 3	*Rapanea myricoides*
	7393	1, 2, 6, 7	*Rapanea myricoides*
	7755	1, 2, 6, 7	*Rapanea myricoides*
	8853	8, 9	*Rapanea myricoides*
	9321	1	*Rapanea myricoides*
	9657	8	*Rapanea myricoides*
	10 824	1, 2, 8	*Rapanea myricoides*
	12 477	1, 20, 21, 22	*Rapanea myricoides*
	14 093	2	*Rapanea juergensenii*
	14 121	2	*Rapanea juergensenii*

These researchers also cross-referenced the card file on folk classification to a file on how the plants were used by the Tzeltal:

In addition to the collection files and cross-indexed files, we drew up 'use' files for several informants. For every Tzeltal plant name in our data, these files include information concerning known utilizations of that plant. For example, if a medicine, what does it cure, what portion(s) of the plant are employed for each use, its wild or cultivated state, where grown and where it occurs, what time of year available and related information.

By this careful process of comparing different types of data, Berlin began to elucidate the structure of Tzeltal plant classification while he was still conducting fieldwork.

1.6.4 The structure of databases

Computers have revolutionized data organization. After entering each piece of information just once in a database, we are able to sort and cross-index the data in as many different orderings as we wish in a matter of seconds. Computers also allow us to organize the data in a way that facilitates statistical analysis, as discussed below.

Whether or not you have a computer, it is efficient to organize your data in the same way as people who use databases. The first step is the define the **fields**, the categories by which you describe people, plants, ecological zones or whatever else you are studying. For example, in a data set on the local people who participate in your study, you may have fields for age, gender, marital status and so forth. For a set of plant collections, you might define the fields as collection number, local plant name, scientific plant name and so on. The number, names and length of all fields is called the **structure** of the database.

The data collected on each separate person or item constitute a **record**. As can be seen in Figure 1.2, the names of the fields are placed across the top of the page,

		FIELDS		
		Name	Age	Gender	Married?
	1	Jack Smith	45	Male	Yes
	2	Clare Smith	42	Female	Yes
	3	Helen Jacobs	53	Female	No
RECORDS	4	Roger Good	35	Male	Yes
	5	Janet Good	36	Female	Yes
	6	Fred Mackey	27	Male	No
	7	George Nutley	38	Male	Yes
	8	Julie Nutley	32	Female	Yes
	9	Bobby Nutley	9	Male	No
	10	Joanne Ashton	21	Female	No
	...	...	...	...	...

Figure 1.2 A sample data set, showing the records, fields and values in an improvised database of project participants.

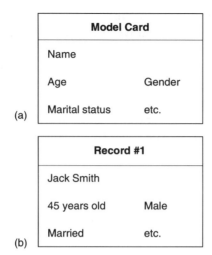

(a)

(b)

Figure 1.3 (a) A model card with the position of each data field marked. (b) An example card, giving values for the first record.

forming columns. Each row corresponds to a separate data record. The shaded area of the table is filled with the **values**, that is, the specific data that correspond to each field and record.

You can make a card file following this format. Each separate card corresponds to a different record and a certain place on each card is designated for each separate field. It is easiest to remember the format if you make a model card, as shown in Figure 1.3a. Each record can then be filled out following the model, as shown in Figure 1.3b.

Whether you are using a computerized database or handwritten notecards, you should carefully conceive the design of the fields, the selection of records and the precise way of writing the values before beginning to enter data. It is best to design a simple structure which is used throughout the project. If you decide that it must be changed during the course of the project, you will have to revise all of the data already recorded, which is often time-consuming and complicated.

Many fieldworkers, grown weary of writing out the same information repeatedly, are tempted to use codes, identification numbers and abbreviations when recording data. In the field for symptoms of illness, for instance, cough becomes 'cgh', sore throat becomes 'srt' and so on. Juan Pérez is converted to informant number '415' and his wife Margarita to '416'. In an encoding of ecological stages of succession, an abandoned field is recorded as 'ESS1' and scrubland as 'ESS2'.

Although it may appear to save time, this practice is likely to increase the amount of mistakes made when entering the data. Errors in coded data are particularly difficult to detect at a later stage. Imagine that you make a list of men and women from your field notes or your card file. If you find 'Margarita Pérez'

15

among the men, you are apt to recognize it as an error. If you find informant number '416', you will have no intuitive idea of whether the person is a man or a woman.

If you do use codes, take them from a recognized reference to which other ethnobotanists have access. For example, when I record species determinations for Sierra Norte plants, I encode the botanical family using three-letter abbreviations that were published by the American botanist W.A. Weber in a recognized botanical journal, Taxon [11]. The codes have the benefit of being **mnemonic** (aiding our memory) because each one is derived from the full name of the plant family. For example, AST makes us think of Asteraceae, POA of Poaceae and so forth.

1.6.5 Protecting data

Many ethnobotanists lose data in one way or another at some point in their career. Carefully prepared plants mold in the humid tropics, field notes are confiscated by suspicious police, information meticulously entered into a field computer is unintentionally erased. Other sources of data disappear with time; photographs deteriorate, voucher specimens are eaten by insects, notes written in non-permanent ink or on poor-quality paper fade.

What can be done to protect these sources of information? There are a few simple rules to follow. Use high-quality materials, including fresh photographic film, acid-free paper and indelible ink pens. Protect your data from damage by light, humidity, insects and other elements. Make extra vouchers of plants, photocopies of field notes, back-ups of computer disks, duplicates of slides and so on. Store the various sets in different locations.

Your data must also be protected from misuse by other people, although this is sometimes difficult to control. There is always a risk that somebody else publishes your original ideas or data before you do, depriving you and your colleagues of deserved credit. The best way to protect yourself against this possibility is to publish your research in recognized journals, in books or even in working papers that are printed by your research institute, non-profit organization or community.

Ironically, by publishing your results for the general public, you are exposing yourself to a second danger – having someone use your data for their personal profit, often in a way that contributes to the impoverishment of local people and the destruction of wildlands. As yet, there are few effective ways of preventing people from using research data for commercial purposes without the consent of participating communities.

The defense of Intellectual Property Rights (IPR) is being actively debated among ethnobotanists. Some explicit guidelines, presented by various individuals and organizations, are further discussed in Chapter 8. Be aware of who is enquiring about your research and what interest they have in the results. Decide whether or not you wish to collaborate with research institutes or laboratories that do not release the results of studies to you personally or to the general public. Support efforts to establish and follow ethical guidelines in research and in the commercial development of research results. Above all, build local capacity to carry out research and to apply the results to community development and conservation where you work.

1.6.6 Analyzing the data

While in the field, it is helpful to make some initial analyses of the data. This is the best way to discover, before too much time has passed, if you are asking the right questions in a meaningful way and if your units of measurement are appropriate. During the course of the research, you might be inspired to ask new questions, to modify your way of collecting data or to improvise new ways of analysis. This flexibility is fine as long as the data collection proceeds systematically.

Analysis should be broad, resulting in an explanation that is valid for the entire data set. In ethnobotany, the most successful analyses are those which can help us understand the use or classification of biological resources across the entire set of specimens collected or categories elicited. For example, let's say you are studying if women in a certain village know more about medicinal plants than men. If you discover that women are more knowledgeable about 15 of the 20 most important medicinal plants, you must explain why the other five plants did not meet your expectations. It is by stating, testing, rejecting and modifying our original impressions that we delve further into our data and refine our analysis.

An important tool for analyzing data is **statistics**, a mathematical way of summarizing and interpreting quantifiable research results [12, 13]. In ethnobotany, we make frequent use of **descriptive statistics** to portray trends in our data, such as '9 of the 12 informants use chamomile tea for stomachache' or '93% of the villagers use oak as a firewood' or 'The average age of the people interviewed is 42.3 years'.

A more complex set of statistics, called **inferential statistics**, is used to provide deeper analyses than those given by descriptive statistics. Inferential measures allow us to make an educated guess about information for an entire population when we have spoken to only a subsample of people in the community. For example, if we have interviewed 50 people in a community of 500, we can use our results to estimate what the other 450 people would say if we asked them the same questions. Some of the most common statistics used in ethnobotany and related fields are the chi-square test, analysis of variance, correlation and regression. In order to use these statistics, you should have a good background in mathematics and access to a computer. Although it is not necessary to know how to derive the mathematical formulae used in these statistics, it is important to understand when to apply each statistical tool and how to interpret the results and explain your conclusions to others. If you choose to employ inferential statistics in your research, consult an introductory textbook that contains a thorough discussion of probabilities and simple explanations of various techniques.

1.6.7 Presenting the data

Scientific writing and public speaking are arts in themselves. Many academics communicate in a technical way meant only to be understood by other colleagues who have an extensive background in the discipline and are already familiar with all of the specialized vocabulary being used.

If you wish for your conclusions to be accessible to a broad range of people, you do not have the luxury of writing and speaking in such a complex way. In the academic world, ethnobotanists seek to convey their ideas to linguists, anthropologists, chemists, botanists and many other colleagues from diverse disciplines. Outside the university, we seek to influence the decisions of government officials and development specialists and to return the results of our studies to communities. Unless we communicate in a style that is both accurate and accessible, we will not reach a large portion of our readers.

How do we go about conveying our ideas in a simple yet precise way? There are many techniques to scientific writing that can be learned by practice and by reading the works of accomplished natural history writers such as Gary Nabhan, an ethnobotanist who has worked in Mexico and the American Southwest [14, 15]. Make your sentences concise, stripped of unnecessary adjectives. Avoid jargon and abbreviations when possible. If used, explain them in simple words the first time they appear. Choose a conversational style, writing down your ideas as you would explain them to a friend or colleague. Ensure that you do not lose scientific rigor in your effort to make the results accessible. The goal is to provide a popular account for the general public while providing the basic data and analyses colleagues seek.

The key to clear writing is revision. Prepare a draft of the text, then read it over to make corrections. Have a friend or colleague make comments on the draft and then revise it another time. Read it out loud to hear how it sounds. Leave it for a day or two and then reread it. A first draft will be full of crossed-out sentences, changed words, added passages and other modifications. The second draft should be cleaner and the final text well polished.

1.7 Visual aids

Images enhance our ability to understand written and oral presentation of project results. **Photographs** – be they close-ups of a flower or wide-angle shots of a landscape – allow us to grasp a complex concept in a single glance. The photographs that turn out clearest in publications are of high contrast, which means that the image has distinct whites and blacks and few intermediate gray tones. The photos should be printed on glossy paper and – like all illustrations – should be accompanied by a concise caption that explains what is depicted.

Color slides enhance oral presentations and can be used to create color or black and white plates in publications. Both slides and black and white prints have added value if you document when and where they are taken. While in the field, some ethnobotanists use a pre-printed form on which they record the type of film used, including the film speed. They write down the date and give a separate number to each roll of film and each exposure. They describe the subject of each photograph as well as relevant features such as the locality, habitat and elevation. When a camera is being used by various members of a research team, the name of the photographer who takes each frame is indicated, allowing credit to be given if the pictures are subsequently published or used in public presentations. After the film is developed, the appropriate number and notes are written on the frames of color slides or on the archival-quality plastic sleeves containing black and white negatives. Photographs which are carefully dated and documented in this way provide an important visual record of changes in local cultures and natural vegetation.

Figures, drawn by hand or with the aid of a computer, allow us to present a simplified version of an image, to highlight its most salient characteristics or to accentuate what we want readers to observe. Instead of including a snapshot of a flower, we can present a line drawing which shows the most important features of the leaves, flowers, fruits and overall habit of a plant. In the place of a picture of a mountain slope, we can provide an ecological profile that shows agricultural zones and the most common types of trees at different elevations.

Tables present data in rows and columns, permitting us to contrast values or examples of related categories. **Graphs** typically relate two dimensions such as quantity, time or membership in a category. Several types of graphs are static, providing us with a 'snapshot' of results at a certain point in time. **Bar graphs** compare the quantity of a single dimension of various related categories or objects – the number of plants found in various use categories or the number of people in different age groups in a community. The height or length of each bar indicates the quantity on a numerical scale which is shown on the side or bottom of the graph. **Pie charts** are aptly named – they are round like pies and are divided into slices of varying size. Each slice indicates the relative proportion or percentage of one category as compared with the others.

Other graphs are dynamic, showing how things change over time or how they change in relation to one another. The **line graph** or **xy graph** permits us to understand how one variable changes in relation to another – how the number of plant names people know is related to how many years they have attended school, for example. The use of tables and these various types of graphs is illustrated in Box 1.3.

1.8 The law of diminishing returns

Because all ethnobotanists work with limited time and money, it is important to decide how much data to compile before we consider our study complete. Do we

Box 1.3 Presenting data with table and graphs – an improvised example

In the chapters that follow, I present the results of several case studies by using tables and graphs. For readers who are unfamiliar with these visual aids, the following improvised example based on a community plant collecting project will demonstrate how to interpret these graphics and to use them when reporting the results of ethnobotanical studies.

Let's say that three members of a community in Jamaica, (a) John Hilton, (b) Mary Jacobs and (c) Phillip Hodges, are chosen by local authorities to make a complete inventory of the flora. After participating in a training course and being equipped with plant presses, dryers and other essential equipment, they all begin to collect plants at the same time.

In order to keep track of their progress, the community supervisor creates a table to record the number of collections made by each collector, as depicted in Table 1.3. The table is split into four columns – one to the left entitled 'Number of months', which is filled with numbers from 1 to 12, and three to the right for the various collectors. At the bottom, there is a caption which explains what is shown in the table. At the end of each month, the supervisor records the values – the cumulative number of collections made by each of the three participants.

Table 1.3 The cumulative number of collections made by three fictitious collectors over a 12-month period

Number of months	John Hilton	Mary Jacobs	Phillip Hodges
1	120	25	5
2	250	60	15
3	420	130	35
4	600	210	70
5	700	310	120
6	750	410	180
7	775	500	240
8	788	580	320
9	794	650	425
10	797	710	540
11	799	760	680
12	800	800	800

The supervisor, who makes a report every three months to the local authorities and the non-profit organization that funds the community inventory, prepares the bar graphs shown in Figure 1.4 to compare the performance of the collectors. As can be observed in the table and these graphs, John Hilton gathered many plants in the first six months of the project. By the end of the first year, all collectors had the same number of collections.

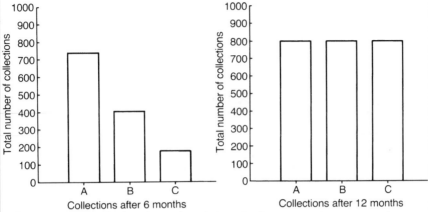

Figure 1.4 Bar graphs that depict the number of collections made by three fictitious collectors after 6 and 12 months of work.

To further illustrate these results from another perspective, she uses pie charts (Figure 1.5) to depict the proportion of total collections made by each participant at six-month intervals. In order to highlight the work of Phillip Hodges (whom she at first thought to be the least efficient of the three collectors) she pulls out or **explodes** the slice of the pie that corresponds to the number of collections he has made.

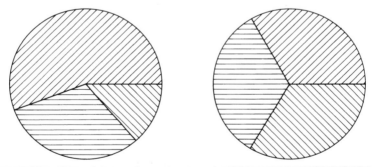

Figure 1.5 Pie charts showing the proportion of collections made by each collector after 6 and 12 months.

At the end of the year, the local authorities ask her to compare the performance of the collectors over the past 12 months not with a 'snapshot' graph of the results, but rather with a graph demonstrating how the number of collections made by each person changed over time. At first, the supervisor depicts this by combining the results in a single figure, called a **stacked-bar graph**. In Figure 1.6, each bar is split into segments that correspond to the number of collections made by each participant over the four intervals of three months. This graph allows us to observe that John Hilton has steadily decreased the amount of collections he makes, Mary Jacobs has made most of her collections between the 3rd and 9th months and Phillip Hodges has rapidly increased the number of collections he makes.

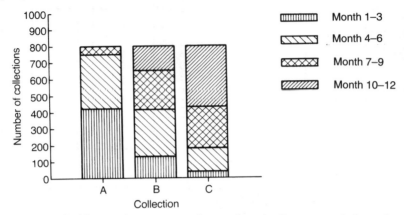

Figure 1.6 Stacked bar graphs comparing the number of collections made by each assistant after 3, 6, 9 and 12 months.

The non-profit organization asks her to predict which collectors would be most productive in the second year. On the xy graph shown in Figure 1.7, the supervisor shows that the collections of John Hilton are leveling off, indicating he will add few additional plants to the inventory. Mary Jacobs is increasing the number of collections she makes per month, but at an ever slower rate. Phillip Hodges is picking up speed in his collecting, indicating that he will be very productive in the second year of the project.

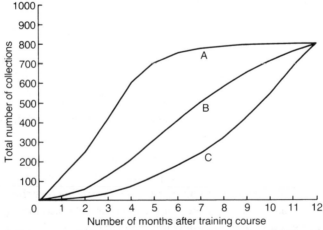

Figure 1.7 An xy graph showing the increase in the number of collections made by each assistant over a 12-month period.

22

The supervisor, curious why each collaborator should collect in such a distinct way, interviews the various participants in the inventory. She discovers that John Hilton, a farmer, had quite a bit of free time in the first part of the year, but later became increasingly busy planting, tending and harvesting his crops. Mary Jacobs is a student whose ability to make collections fluctuates with the demands of her course work. Phillip Hodges is a health promoter who was very busy with other duties at the beginning of the ethnobotanical project, but managed to make more and more specimens as time went on. Eight months into the year, the health project ended and he decided to dedicate himself full-time to the ethnobotanical work.

need to collect 100, 1000 or 10 000 plants? Shall we stay in each village for a week, a month or a year? Will we have enough information after talking to five, ten or 100 people? Should we set up plots of 0.05, 0.5 or 1 hectare?

The efficiency of our data collection is governed by the law of diminishing returns – we gain a decreasing amount of new information from each interview or plant collection that we make. We may learn much from the first person to whom we talk. The second interview will give us some new information, but some of what the first informant told us will probably be repeated. The third respondent will probably give us even less original data. Eventually, we will find that each new person we talk with tells us things that we have already heard from others. Hearing the same thing several times helps us to confirm our data and noting disagreement between informants allows us to gauge cultural variation in the community. But at some point, it simply becomes inefficient and uninteresting to keep asking the same questions over and over again.

The same principle applies to plant and animal collections. It is good to collect the same species with several different informants, but unless we are particularly interested in measuring variation in knowledge, we need to draw the line after gathering a certain number of specimens. In several examples in the chapters that follow, I discuss ways of deciding when 'enough is enough'.

1.9 Hypothesis testing and theory

Wade Davis, a Canadian ethnobotanist, has explained why ethnobotany still depends a great deal on extensive data and specimen gathering [16]:

... critics of the practice of ethnobotany usually overlook two important considerations. First, the act of compiling raw information provides the foundation of any natural science and without a basic inventory theoretical formulations are not possible. Second, ethnobotany remains on one level what it has always been – a science of discovery. Its contributions to the welfare of mankind have not been trivial.... Today, in an era marked by the massive destruction of diversity not only of plants and animals, but of human cultures as well, basic plant exploration remains a vital and essential contribution of the ethnobotanist.

Although compiling raw data is a fundamental activity, our ability to contribute to human welfare and scientific discovery is directly related to our formulation of general principles (based on many individual observations) that are the foundation of ethnobotany. Testing our tentative impressions of how people interact with their natural environment adds rigor to the way we collect data and allows other researchers to evaluate our methodology as they continue their own quest for knowledge.

For these reasons, research should be guided by **theories** and **hypotheses**. A theory is a general model or set of principles that describe natural and social phenomena. From theory, we derive hypotheses – the tentative assumptions which can be explored through observation and empirical testing and which lead us on to more sophisticated ways of collecting data and deeper interpretations of research results. When a hypothesis is supported by our analysis of the data, the theory stands. When a hypothesis is rejected through discovery of information that contradicts our tentative conclusion, the theory must be modified or discarded.

For example, there is a theory that people in a community generally use the same names for the same botanical species because this allows them to communicate about plants that are important in their everyday life and to pass along this knowledge to future generations. A testable hypothesis derived from this theory is that speakers of the same language will give identical names to plants collected in their community.

Let's say we actually ask a broad range of people in a given village to identify a set of plants. We discover that culturally important species are consistently named whereas those with minor uses are called by various names. This would require us to modify the general principle of linguistic behavior that we had first formulated, perhaps restating it as 'people in a community tend to give the same names to biological organisms but their consistency increases with the cultural importance of the individual plants and animals'.

Oliver Phillips, a British biologist whose work is further described in Chapters 5 and 6, has noted that there is an over-emphasis on data compilation and a relative scarcity of hypothesis-testing in ethnobotany [3, 4]. He describes several areas in which quantitative approaches and statistical hypothesis-testing have been employed in recent years, which he takes as evidence of a new trend towards scientific rigor in ethnobotany.

In his own studies of how Mestizo agriculturalists use plants in southern Peru, Phillips uses statistics to compare the cultural importance of different plant families, the ethnobotanical knowledge of local participants and the contribution of various ecological and botanical factors to determining the utility of individual species in the eyes of local people. He refers to studies of other colleagues which

24

focus on diverse issues such as evaluating the importance of plants or vegetation types to one or more ethnic groups, establishing the relative cultural significance of various useful plants and testing a model of the origin of medicinal plant use.

A common element in all of these questions is that they can be posed in the form of a hypothesis. Although there can be no conclusive proof for such ideas, it is often possible to find evidence that allows us to reject alternative interpretations, thus adding strength to our argument. In other words, scientists make progress not by proving hypotheses but by falsifying them.

Statisticians have a rigorous way of posing, accepting and rejecting hypothesis. Their strategy consists in proposing that there is no statistical difference between two or more measurements or observations. This is referred to as the **null hypothesis**, a term derived from the idea that there is *no* significant divergence or pattern in the data. For example, you are stating a null hypothesis when you make statements such as 'there is no significant difference in what men and women know about primary forest trees' or 'plants chosen randomly are as physiologically active in pharmacological screens as those selected because they are used medicinally by local people' or 'there is no difference in the cultural utility of different plant families'.

For each of these statements there is an **alternative hypothesis** which is automatically accepted if the null hypothesis is rejected. If you have the statistical evidence to reject the assumption that there is no difference in what men and women know about primary forest trees in a community, you automatically accept the alternative hypothesis that there are statistically significant gender-based patterns of knowledge.

After collecting the appropriate data, inferential statistics are employed to analyze the validity of the hypothesis. The results of these statistics are given as a probability, represented by the letter p. Probability is expressed in decimal form as a number that ranges between 0.00 and 1.00. The higher the number, the more probable is the result. For example, a p of 0.53 means the result is more likely than a p of 0.28. In general, researchers reject a null hypothesis if it is shown to have a p of less than 0.05 and feel they have an increasingly strong case as p becomes less than 0.01. This is often expressed by saying 'the result is significant at the 5% (or 1%) level'.

As you review the techniques presented in the following chapters, pay attention to the various hypotheses that have been proposed by ethnobotanists and the way they test their ideas. In your own work, seek to state your tentative conclusions in the form of hypotheses that can be further tested in the field or laboratory.

2

Botany

Figure 2.1 Dennis Soibi and Meliden Giking of the Kinabalu Ethnobotany Project arranging plant presses in a dryer in Kinabalu Park headquarters in Sabah, Malaysia.

2.1 Collecting and identifying plants

Among the most basic skills in ethnobotany is plant collecting. Collections are valuable because they serve as **voucher specimens**, permanent records of the plants that are known by a certain community, as further explained in Box 2.1. They also function as **specimens for determination**, allowing plant taxonomists to identify the family, genus and species of a collection. In some studies, they act as **reference specimens**, the samples used in naming, sorting and other tasks carried out with local participants.

Box 2.1 The value of ethnobotanical voucher specimens

In a presentation to the 15th annual conference of the Society for Ethnobiology, held at the Smithsonian Institution in 1992, Eugene Hunn noted [17]:

> ... the voucher specimen is the link between two bodies of information, that of Western biological science and that of the ethnoscience of the native culture the ethnobiologist seeks to document. For example, Sahaptin-speaking Indians of the Columbia Plateau employ a plant they call *chalu'ksh* for a variety of purposes, nutritional, medicinal and as a fish poison. This fact remains an ethnographic particularity, however, until it can be established that *chalu'ksh* means *Lomatium dissectum* (*Apiaceae*). On the basis of this equation it is possible to compare a segment of Sahaptin ethnoscientific knowledge with a corresponding segment of Western botanical systematics, phenology, ecology and pharmacology. This equation also makes possible comparisons with the ethnoscientific traditions of other cultures within the range of this species. The resulting synthesis is of greater value than the sum of its parts, the disconnected bits of ethnographic detail we would otherwise have to deal with.

Hunn draws his definition of a voucher specimen from another colleague, Robert Bye, who has carried out extensive research on market ethnobotany in Mexico and on the plants used by the Tarahumara, an indigenous group of northern Mexico. Bye has pointed out that a voucher specimen facilitates the identification of the plants and animals encountered during research and permits colleagues to review the results of the study. To serve this purpose, the material chosen as a voucher must: (1) have diagnostic characters which are easily recognizable; (2) be preserved and maintained in good condition; (3) be thoroughly documented by taking field notes on the collection locality and the appearance of the organism as well as its classification and use by local people and (4) be readily accessible in an institution that is clearly identified in research reports and publications.

Although many ethnobotanists use the words 'collections' and 's
interchangeably to refer to the plants they gather, these and other t
specific meanings for professional botanists. A **collection**, designated b
collection number, refers to a set of plants pertaining to one species collected at
the same time in one locality. Each specimen in a numbered set is called a
duplicate. Once mounted and placed in the herbarium, the collections are referred
to as **herbarium specimens**. Herbarium specimens are usually contained on one
sheet of mounting paper, but some bulky plants such as palms may be divided to
make two, three or more sheets.

A good quality herbarium specimen contains a representative sample of the
plant, including stems, leaves, roots, flowers, fruits and other plant parts which are
characteristic of the species. Whenever possible, whole plants or entire branches
are included so that the overall architecture of the plant can be observed. The
specimen is pressed flat and dried in the field.

It is later mounted in the herbarium on sheets of high-quality paper which
measure approximately 28.5×42 cm. A label in the lower right hand corner gives
the date, locality, collector and collection number of the specimen as well as notes
on morphological features of the living plant that are not evident in the dried
sample. There may be an annotation label which provides the correct scientific
name of the plant. An ethnobotanical label with information on local names and
uses is sometimes attached to the specimen. An example herbarium sheet is shown
in Figure 2.2.

How do we go about making good collections? Although plant collecting is
often considered a mundane activity, it is an art form in itself. The following
sections cover the seven basics: choosing a locality, collecting the plants, pressing,
drying, keeping a field notebook, labelling and distributing the specimens for
identification.

2.1.1 Selecting a locality and population of plants

A day of plant collecting begins by selecting a site and the plant populations to be
collected. For each locality, observe the dominant vegetation, soil type and
relative amount of sun or shade. Gauge the slope of the land and which direction
(east, west, etc.) it faces. Ask for the local name of each site and note the altitude
and the distance from a known landmark. In order to do this accurately, bring
along the following tools: a compass, which indicates direction, a topographical
map, or another map which shows the rivers, human settlements and other
geographical aspects of the terrain, and an **altimeter**, which measures elevation
above sea level. Alternatively, the altitude may be estimated from the topographical map.

Choose a site where collections will have little impact on the vegetation and
on the productive activities of local people. When collecting in conservation areas,
take care not to remove plants that add to the beauty of well-traveled paths or

Figure 2.2 An herbarium specimen of *Tagetes lucida* Kunth collected in the Sierra Norte of Oaxaca. The sheet has a standard label, an ethnobotanical label and an herbarium accession number.

other areas much visited. No matter where you collect, be careful not to diminish populations of rare plants [18]. If you are collecting a large number of duplicates or bulk material, use techniques – further described in Chapter 3 – that do not kill the plants or endanger the survival of the local stand of the species.

When collecting near communities, be aware of which plants are being managed or cultivated by the residents. Some species may be protected because they have religious significance or grow in sacred sites. Other plants may be left because it is socially unacceptable to gather them. For example, some local collectors in the Sierra Norte are hesitant to collect jimson weed (*Datura* spp.) because it is reputed to be used in witchcraft. If you gather species like this in an obvious way, people may be suspicious of your intentions. For all of these reasons, ask local people for permission before collecting samples in their communities and follow their wishes.

At times, human activity provides opportunities for collecting. Just after a forest is cut during a logging operation or to cultivate the land, you can make collections from the canopies and trunks of felled trees, where difficult-to-reach flowers, foliage, epiphytes and vines can be found. Collecting along the edge of the primary forests, in gaps or near tree-falls is also efficient, because foliage, flowers and fruit of woody species are found closer to the ground. Because of the increased light found in these areas, some species can be found in fruit or flower during a collecting trip, even if they are found only in sterile condition deeper in the forest. Agricultural activities have much the same effect. They keep the vegetation in various stages of succession, providing a series of niches in which diverse species might be found.

The thoroughness of collecting is determined by the goal of the study and the diversity of the local vegetation. If you are studying variation in classification, you will want to sample various populations of the same species that different local people show you. If collecting species that exhibit morphological variation, you should document at least all of the varieties named locally. This is particularly critical when making an inventory of domesticated plants, which are often categorized into numerous folk varietal categories. Over the course of a long-term project, you should collect widely enough to ensure that your specimens represent local floristic diversity, including species from the various vegetational formations, soil types and climatic zones.

Pay close attention to any variation exhibited by plants that belong to the same folk category. Remember that there is not always a one-to-one correspondence between folk categories and scientific species. In order to document the entire range of plants included in a folk generic, make a number of collections with various informants.

2.1.2 Collecting the specimens

Once the locality and plant population have been chosen, you can select the portion of each plant to collect. Keep two points in mind: (1) the material selected must be flattened in standard sized newspaper (about 30 cm wide × 45 cm long, when folded) and (2) it must represent as well as possible the morphological features of the species. These two goals may appear contradictory, since squashing a plant between newsprint often reduces it to a poor vestige of how it looked when growing. Yet with a little experience, you can prepare specimens that are identifiable, appealing to the eye and useful for further field and herbarium research, which are the criteria that professional botanists consider important.

Table 2.1 Checklist of collecting, pressing and drying equipment for the field and drying area

Field	Drying area
✓ Compass, maps and altimeter	✓ Plant press, including press ends, straps,
✓ Field notebooks and permanent ink pens	cardboard or aluminum corrugates and felt
✓ Cutting tools and sheaths, including clippers,	blotters
knives and machetes	✓ Plant dryer, including frame, screen and
✓ Plant digger or trowel	heating elements
✓ Collecting bags	✓ Butane gas tanks, light bulbs or kerosene for
✓ Jeweller's tags or paper labels	the heating element
✓ Field press, including newspaper and cardboards	✓ Permanent markers
✓ Collecting poles with clippers and saw head	✓ Newspapers
✓ Climbing equipment	✓ Paper and plastic bags of various sizes for
✓ Measuring tapes, clinometer	storage of bulky plant parts
✓ Camera and color slide or black and white film	✓ Pre-printed collection book
✓ Leather gloves	✓ Alcohol or other preserving fluids
	✓ Jars and bottles for spirit collection

Pressed plant collections must document the overall aspect of the plant. They are not intended to show which plant part is used, how it is prepared or with which other plants it grows. For example, if you are trying to discover the scientific name of a medicinal tuber, you must collect not only the root but also the foliage, flowers and fruits. If the root is combined in a remedy with other plants, you must make a separate numbered collection for each species.

There are a few basic tools you should carry and a few rules to follow when collecting plants. An assortment of equipment and materials for fieldwork and drying of specimens is listed in Table 2.1. The diameter of woody plants is measured using a tape measure and the height estimated by sight or with a clinometer. Various cutting tools – including a **knife, plant clippers** and whatever local instruments are used for clearing vegetation – aid in making paths, pruning branches and slicing fruits, roots and other thick plant parts. To dig up the roots of herbaceous plants, bring along a **plant digger** or **trowel. Extendable collecting poles,** armed with clippers and saw head, aid in reaching branches and vines from 2–10 m above the ground. The **clipper and saw head** can be carried to the field and attached to a pole cut in the forest if you wish to avoid being encumbered by aluminum poles. Alternatively, a forked pole can be used to twist off branches and guide them to the ground. Afterwards, the collection is pruned to fit in the newspaper. If tall trees or vines are to be collected at a height of greater than 10 m, **climbing equipment** can allow you to scale tree trunks to reach high branches near the canopy.

Collecting bags let you amass a large amount of plants for later pressing. The plants are cut or folded to the size of the plant press and all individuals of the same species placed together in the bag. Alternatively, you may use a **field press,** which consists of two wooden ends, cardboards, newspapers and straps. Although less efficient than collecting in bags, field pressing prevents damage or wilting of specimens and ensures that you have a sufficient amount of material for the number of duplicates you wish to collect. Both techniques can be employed, placing fragile specimens in the field press and more robust plants in a bag.

Some botanists use small, stringed labels called **jeweller's tags** to put the collection number on each separate plant or plant part before it is placed in the collecting bag or field press. This ensures that collections of different species do not become mixed and that any segments of the specimen that fall out of the paper during transport or handling can be returned to the proper set of duplicates.

A **field notebook** is brought along to record site notes, plant measurements, observations on surrounding vegetation and other pertinent details which are further described below. Whether pre-printed or blank, it should be of sufficient quality to hold together under field conditions. It should have a flexible but durable binding as well as pages that do not stick together when they become moist. Collectors who deposit their complete notes in an herbarium along with their collections opt for high-quality notebooks containing waterproof acid-free paper that will last indefinitely. Other botanists use standard notebooks, but transfer all of their notes to computerized databases, making a permanent record. Carry along plenty of permanent-ink marking pens. Notes written with water-soluble ink tend to run, especially in humid forests, and those written in pencil can be hard to read and photocopy.

Collect as many plant parts of each species as possible. If possible, specimens should have mature flowers, fruits or both. Herbaceous plants, including grasses and ferns, should be collected with roots. When a plant has compound leaves, include several entire leaves on the branch or stem in order to show whether they are opposite, alternate or attached in another way. For woody plants, include portions of the branch large enough to show the pattern of leaf attachment and branching. Some botanists collect a piece of wood and bark when collecting trees, a technique described later in this chapter.

The plant parts selected should be representative of the morphological variation in the plants that you see in nature. Avoid the temptation of choosing plant parts just because they fit easily in the newspaper, especially if large leaves, flowering stalks or fruits are characteristic of the species you are collecting. Botanists typically gather healthy, whole specimens, but will include some diseased parts if they appear to be typical of the species in the collection area.

Collect various developmental stages of the plants. Immature leaves, young branches, stump sprouts, seedlings and other less commonly collected parts enrich our knowledge of the plant's morphology. They are particularly desirable if they correspond to the portions of the plant used by local people. If collected, they should complement, but never replace, mature leaves, twigs, flowers and fruits. When possible, collect extra flowers and fruits to add to each specimen, because taxonomists depend heavily on this material when identifying plants. If you must collect a sterile specimen, make sure that you have good quality vegetative material and keep an eye out for fertile material of the same species on subsequent collecting trips. Box 2.2 explores the controversy over making sterile specimens in ethnobotanical projects.

Ensure that you have collected enough material for all the duplicate specimens that you wish to press. If you are planning to make a field reference collection, as described later in this chapter, select material for a small specimen that will fit on a standard piece of paper, approximately 20 by 25 cm, and which shows the morphological features local people and botanists use when they identify the plant.

Box 2.2 The controversy over sterile collections

A major point of contention between plant taxonomists, ecologists and ethnobotanists is the collection of sterile specimens. All researchers prefer to collect plants with flowers, fruit or both, because they are easier to identify than sterile material. Nevertheless, ethnobotanists and ecologists are sometimes constrained to make collections that consist of a branch with a few leaves, which is all they can find when they are collecting with a particular local person or during a specific season. As an ethnobotanical voucher, a sterile collection that can be identified to botanical family or genus is much better than no collection at all.

Sterile collections are a burden to taxonomists, because they are often difficult to identify, contribute little to taxonomic studies and take up space in herbarium cabinets that could otherwise be used to store fertile specimens. It costs as much to curate a poor quality specimen as a good one. Because most herbaria have limited resources, they must make a selection of the most valuable material to mount and file in the permanent collections.

There are several solutions to this dilemma. If collecting will be done over a period of several years, ethnobotanists can collect only fertile material and constantly seek species that have not yet been sampled. In this way, sterile material can be avoided altogether. If sterile collections must be made, the species should be collected in flower or fruit at a later date, whenever possible. In the later stages of a long-term collecting project, researchers can give special attention to finding good specimens of plants previously found sterile.

Just as ethnobotanists should make an effort to collect fertile specimens, taxonomists should understand the importance of collecting sterile material, especially at the beginning of a long-term project. An example from the Sierra Norte of Oaxaca illustrates how some sterile specimens bring to our attention unique botanical species which can be collected in fruit and flower at a later stage of fieldwork.

Ticondendron incognitum Gómez-Laurito and Gómez P. is a cloud forest tree that was named in 1989 from collections made in Panama and Costa Rica. Because it is very different from other botanical species, it was named as the type species of a new family in 1990. In 1987, two local collectors from the Sierra Norte – José Rivera Reyes and Ricardo López Luna – collected it in sterile condition in Totontepec and La Esperanza. Because of the possibility that the collections might represent a disjunct Mexican population of *Ticondendraceae*, José and Ricardo searched for the tree in fertile condition. As shown in Table 2.2, some of the resulting collections had fruit, enabling us to verify the tentative identification and provide ethnobotanical and ecological data on the little known Mexican populations of this species.

Table 2.2 Various collections of *Ticondendron incognitum* Gómez-Laurito and Gómez P. made by local collectors in the Sierra Norte of Oaxaca, Mexico (JRR, José Rivera Reyes; RLL, Ricardo López Luna).

Collection number	Condition	Date	Locality
JRR 0440	Sterile	12 April 1987	Totontepec
RLL 0049	Sterile	30 September 1987	La Esperanza
JRR 0721	Sterile	11 March 1990	Totontepec
RLL 0656	Fruiting	8 July 1990	La Esperanza
RLL 0694	Old fruits under tree	29 April 1991	La Esperanza

In general, taxonomists recommend that several duplicates be made of each collection number, which allows specimens to be sent for identification and ensures that a voucher will continue to exist even if one duplicate is accidentally lost or destroyed. At times, only a few individuals of a certain species are found. Keep an eye out for more plants, but if none is available it is better to collect one sheet than none, because these single collections increase the breadth of an ethnofloristic inventory.

Let me give an example from Oaxaca. After reviewing the first several hundred collections in the Sierra Norte communities of Comaltepec and Totontepec, I discovered that some plants detected in one place had not shown up in collections from the other. For example, an important source of emergency food for the Mixe is the vegetative base of a shrubby fern, *Marattia weinmannifolia* Liebm. Several collections had been made by Mixe collectors in the cloud forests of Totontepec, but none by Chinantec collectors in the cloud forest of Comaltepec. When reviewing ferns with Ricardo López Luna, I discovered that he had not collected the *Marattia* because it was very scarce in the forests around La Esperanza, where he was collecting. I asked him to gather enough material for a single sheet. His number RLL643, collected on the 25th of April 1990, documents that the fern is also used by the Chinantec people of Comaltepec.

2.1.3 Keeping a field notebook

There are several ways to keep a field notebook, but one golden rule to follow: always take notes in the field while making plant collections rather than relying on your memory at the end of a day or trip. Many botanists prefer to record their notes in a blank notebook, guided by a mental checklist of the observations they wish to include. Others use a pre-printed notebook, which ensures that field data are collected in a certain order and no details are forgotten. Still others write notes on tags which are attached to the plants in the field or on the newspaper in which the collections are placed. These remarks are then transferred to a notebook at the end of the day, when putting the plants in the press.

In addition to standard collection information, ethnobotanists give special attention to recording notes on the uses, local names and other cultural information regarding the plants. Many opt for a pre-printed collection notebook, especially when numerous collectors are collaborating on the project. Figure 2.3 shows a translation of one page of the pre-printed ethnobotanical notebook that is used in Totontepec. Each page is divided into two sections, one that contains the information that plant taxonomists require for label-making and floristic databases and the other for ethnobotanical information. The collector's name, collection number and date are given at the bottom of the page. The various entries are in the same order as the corresponding fields in the computerized databases used to prepare specimen labels and to store ethnobotanical notes.

Carbon paper is placed between successive pages of the notebook to make a copy of each collection sheet. The original is left in the notebook as a permanent record of the collection and the carbon copy is placed in the corresponding set of duplicates when they are removed from the plant press after drying. All information from the carbon copy is entered into a database when the collections are processed for labelling and shipping. Note that there is limited space for recording information about the name, use and preparation of the collections. Additional details of local botanical knowledge can be gained through participant observation, interviews and sorting tasks carried out in the community, as described in Chapter 4.

The way of writing a field notebook will vary according to the goals of the project and personal preference. A pre-printed notebook can be prepared in a participatory way with local collaborators. For example, a group of park guards, researchers and community members designed the ethnobotany collection notebook for the Beni Ethnoecological Project, which is taking place in northern Bolivia. This ensured that the resulting ethnobotanical data will serve the various purposes of the participants in the initiative.

Although it is not possible to dictate the exact structure, some basic information

Ethnobotanical Collections of Totontepec

Community _____ Specific locality _____

Vegetation type: Climate: Soil:

Lifeform: herb   bush   tree   grass   vine

Other, specify _____

If tree, bush, or vine, Height _____ Diameter _____

Color of the flower _____ of the fruit _____

Other notes on the plant's appearance _____

Flowering season _____ Fruiting season _____

_____

Mixe name _____

Translation _____

Use _____

Preparation _____

Use _____

Preparation _____

Other notes on the use and preparation _____

Who gave the information _____

_____

Collector _____ No. _____ Date _____

Figure 2.3 A sample page from the pre-printed ethnobotanical notebook employed in Totontepec.

is obligatory. For example, when botanists start each day of collecting, they begin their note-taking by writing the **date** at the top of the page. For each site in which they collect, they record the following information:

- The **locality** where the collection was made, including all important political divisions that would aid in locating the site on a map or during a subsequent visit to the field site. List the country first, followed by the state or province, county or district and the closest community. When possible, the exact location should be specified, noting the distance (preferably in kilometers) and the direction (north, east, south, west and so on) from a permanent recognizable landmark. Where maps and instruments permit, longitude and latitude should be calculated to the nearest minute and altitude rounded to the nearest 50 m. Describe the type and successional stage of the vegetation. Note if the plants are growing in any particular ecological niche, such as along a river bank, on tree trunks or in a valley bottom. For example, a collection locality near the village of Totontepec could be given as:

Mexico: State of Oaxaca: Municipality of Totontepec: 2 km north of Totontepec along the road to Amatepec. 2000 m above sea level. In disturbed vegetation along the roadside near secondary cloud forest.

- The **environmental aspects** of the collection site. Give brief descriptions of soil color and composition, exposure to the sun, the slope of the land and other features. For example,

    Growing in yellowish, clay soil in a sunny opening along a steep slope of northern exposure.

At each site, one or more species are collected. For each collection, the following information should be noted:

- The **collection number**, a unique identification number for each different population of each species you collect. Most botanists begin at 1 and number consecutively throughout their life. They recommend that you not start a new series of numbers for each project or for each new year that you collect, because this will create confusion when other researchers are citing your collections or searching for them in herbaria. Nevertheless, some researchers employ a yearly system, beginning each collection number with the first two digits of the year, such as 94-1, 94-2 and so on for plants collected in 1994.
- The **identification of the plant**, including when possible the botanical family, genus and species. When working in a new region, this space can be left blank or incomplete for the plants you do not know. As you or collaborating botanists provide determinations, these can be entered in the notebook or directly in a computerized database compiled from the notes.
- The **appearance** and **abundance** of the plant. Remember that other people will see only the plant fragment that you have pressed and dried. They must reconstruct other elements of the plant's morphology and ecology from label information, including the size, shape and lifeform of the plant, where it grows, how common it is and what other plants grow near it. Use standard lifeform terms – tree, shrub, vine, herb, grass, fern and so on – to describe the habit. Note whether the plant is abundant, common, scarce or rare. Include a short description of characteristic aspects of plant morphology that are not evident on the dried specimen, such as flower and fruit color, scent, sap color and consistency, color and texture of the bark, flavor of the fruit and so forth. For example, a collection of a wild variety of the scarlet runner bean might be written as follows:

    A common 2-m scandent vine, climbing on neighboring shrubs and erect herbs, with red flowers and purple-spotted, green fruits.

- **Information about complementary collections**. Depending on the goals of the project, you may collect live specimens, seeds for a germplasm bank, samples for phytochemical analysis, fruits or flowers preserved in alcohol, wood for anatomical study or cultural artefacts made from the plant. There may also be photographs or field drawings of the plant and its habitat. Make a note of these complementary materials in your notebook, giving important details, such as the kind of preservative solution used or the way phytochemical samples were prepared.

2.1.4 Recording ethnobotanical information

After recording the standard information required by botanists, basic data on the local knowledge of the plants are noted. The amount of information you record will depend on the breadth of the ethnofloristic inventory you are making. If you discuss each plant for an hour and record pages of notes, you will not achieve an extensive collection, especially if you have a short time in the field. Yet if your preliminary notes are very sketchy and you do not have the opportunity to conduct detailed interviews at a later date, your data will be of limited usefulness to linguists, anthropologists, pharmacognosists and to the community itself.

Even if you will be gathering detailed ethnographic data during a later phase of the project, record some initial observations from at least one informant as you make plant collections. If you are working for the first time in a community, the questions you ask will tend to be naive and your ability to understand and verify the answers will be limited by your lack of experience with the local language and culture. Whenever possible, the responses should be verified by eliciting information with a number of different techniques as described in Chapter 4.

Consider recording the following data in your notebook:

- The **local plant name** used by the informant. If you are not able to transcribe the local language accurately, you might choose to tape record the names and later ask a linguist familiar with the language to transcribe them. Ask for a word-for-word translation of the name in the local contact language and for an explanation of its meaning, that is, why the plant is called by that name.

  For example, *Salmea scandens* (L.) DC. is a common vine in the Sierra Norte of Oaxaca. José Rivera Reyes, a collector in Totontepec, calls it **niiv aa'ts** in Mixe. When we were reviewing a collection of the species, he explained to me in Spanish that **niiv** means *chile* (chili pepper) and that **aa'ts** means *bejuco* (vine). The Salmea is called 'chile vine' because it is the source of a condiment reminiscent of chili peppers.

  In some cases, informants will say that a particular name refers only to the plant and that it has no other meaning. For example, the Mixe name **xoj** is the proper name for oaks and cannot be further interpreted. Remember that requesting the translation and meaning of the name is only the first step in understanding folk botanical nomenclature. When possible, these first questions should be followed by a more rigorous analysis as described in Chapter 7.

- The **lifeform,** or global category into which the plant is placed by local people.

The Mixe assistants record the lifeform of each collection as **kųp** (tree), **ojts** (herb), **aa'ts** (vine), **tsoots** (grass) or another category. The lifeform to which a particular generic belongs can be verified by speaking with other informants or by carrying out sorting tasks, as explained in Chapter 4.

- The **characteristics** that local people use to identify the plant. Although people often recognize plants at a glance, they are usually able to point out some features that distinguish one folk category from another. Observation of these key features may depend on any one of the five senses – the taste of the berries, the feeling of the cutting edge of a grass, the color of the flowers, the smell of the crushed leaves or even the sound of the foliage rustling in the wind.

- The **local uses and preparation** of the plant. The amount of detail recorded will vary with the approach of each ethnobotanist, the time spent in the field and the other opportunities to collect data on folk knowledge of plants. When collecting large numbers of plants in a community where in-depth interviews will be made, it is sufficient to make brief notes while collecting, such as 'taken in tea for stomachache'. These remarks can be explored in greater detail during longer conversations with local people after the collecting is finished. When no additional interaction is foreseen, it is important to collect as much information as possible when the plant is collected.

- **Data on the person** who gave the information. You should record the name, age, gender, place of residence and occupation of people who work with you. Note whether or not the informant has an occupation which entails a special knowledge of plants, such as being a traditional curer, carpenter or midwife. In-depth studies require selection of a representative sample of local participants, as described in the section on anthropology.

You may wish to ask additional questions about the folk appraisal of the ecological features of the collecting site, including the type of soil, climate and successional stage of the vegetation. You can also record the toponym, or local geographic name of the site and the human ecological zone to which it pertains. For example, José Rivera Reyes made many collections in open, relatively flat agricultural lands below Totontepec. This site, which is called by the proper name **pa'tm** in Mixe, is in the 'temperate zone' (**mukojk an it**) and is characterized by its 'black soil' (**yąk naax**). Most of the land has been dedicated to cultivated fields (**kam**) but some has been left to fallow (**kam_tajk**) and other parts have reverted to shrubby vegetation (**peji kam**). Every time that José made a collection in **pa'tm**, he recorded the specific successional stage in which he found the plant as well as the climate and soil type. These data parallel and enrich the geographical notes on a field site that botanists usually record.

2.1.5 Composing and arranging the specimens

At the end of a day's collecting, the plants are taken to a drying area for processing. There should be a large store of old newspaper on hand to press specimens which have been stored in bags or to use as replacements for torn or wet newspaper which contain field-pressed specimens. If you have been collecting in the rain or in humid areas, the plants may be spread out on the floor in separate piles, each corresponding to a different species. When the surface moisture has dried, proceed to press them in dry newspaper. When working in dry areas, wilting can be prevented by wrapping plants in wet newspaper in the field or by trimming the stems and placing the plants in water in the drying room.

In tropical humid areas, you may be collecting plants in a remote location where you do not have access to a dryer. Under these conditions, the pressed plants may be preserved in alcohol – as described in Box 2.3 – which will keep them free from mold or decay until they can be dried.

Box 2.3 Preserving pressed plants in alcohol

Preserving pressed plants in alcohol requires less equipment in the field than normal drying of specimens, because presses, cardboards, blotters and the dryer may be left at home. It also saves time that would have been spent in drying specimens in the field, thus increasing the number of specimens that can be gathered during a field trip. Freed from the necessity of being near a dryer, botanists can venture into remote parts of the forest for several days at a time, reaching localities where nobody has yet collected.

Nevertheless, in ethnobotanical projects the benefits of preserving plants in alcohol are outweighed by the disadvantages. The plants and newspapers sodden in alcohol are a burden to carry. The plants cannot be handled again until they have been thoroughly dried, which means they cannot be shown to additional informants in the field to verify local names and uses. If the plants are to be carried out of the country for final processing, duplicates cannot be left immediately in the host country, but must be mailed later. The alcohol treatment creates unattractive, blackened and brittle specimens that are not much appreciated by either botanists or local people. Furthermore, the plants cannot be used later for chemical analyses.

When absolutely necessary, the field-pressed plants can be preserved by employing the following method, paraphrased from *Realisation d'un herbier tropical* – a document written by M. Hoff and G. Cremers which describes botanical collecting and the construction of an herbarium in Cayenne, French Guyana – and from a manuscript about plant collecting techniques written by Ron Liesner, botanist at the Missouri Botanical Garden in the United States [19, 20].

The first step is to number the newspapers with a marker which is not soluble in the preserving solution and to place the pressed specimens in a leak-proof plastic bag. For each 15–20 cm tall stack of plants, pour in about one liter of 60% or 70% isopropanol or ethanol. If laboratory grade alcohol is not available, locally distilled drinking alcohol is an acceptable substitute. Some botanists add a small amount of acetic acid (in the form of vinegar, if necessary) and formol.

After the preserving solution is added, the bag is closed tightly and is then turned several times to distribute the alcohol evenly. It is best to store the bags flat, then to turn them the next day and again the following day. This ensures that the alcohol will thoroughly penetrate the bundle. If sealed in several layers of plastic bags which do not have holes, the plants can be stored for several weeks or months before drying them. When storing plants in this way, add alcohol from time to time to replace that which has evaporated.

After reaching a place where drying facilities are available, the preserved plants can be placed in a press and dried normally. Although it can be difficult to rearrange the plants, any creases or folds in the newspaper should be straightened out, avoiding difficulties in opening them once they are dried. Take particular care when removing the plants from the press, because they will be more brittle than plants which have been dried immediately after collecting.

Of the many tips on how to arrange good specimens, I describe only several important ones here. As with all skills, plant collecting is better learned by doing than reading. If possible, work alongside an experienced botanist in the field so that you can learn firsthand. Various suggestions are illustrated in Figure 2.4.

Each specimen should be as complete as possible, including a sample of all the plant parts collected. Do not put all the flowering and fruiting branches in one or two duplicates, leaving the others sterile. Select a sufficient amount of material to fill the newspaper, remembering that many plants shrink upon drying. The plants should not hang out beyond the edges of the newspaper nor should much empty space be left inside it. When plants are small, many are needed to fill the sheet, whereas large plants must be cut down to size.

Do not allow leaves, flowers and other plant parts to overlap, because they will dry more slowly (increasing the risk of molding or color loss) and they will be difficult to study in the dried specimen. If some overlapping leaves are cut off, leave their stem intact to show how they were attached to the twig. When the plant is large or bulky, the various plant parts can be placed in separate sheets of newspaper. Large leaves should be folded in such a way as to show the tip, base and overall form of the leaf.

Plants must be completely flattened and no part of the specimen should be

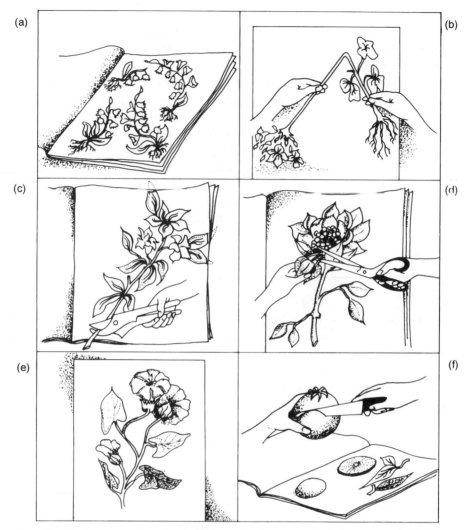

Figure 2.4 Various suggestions for arranging plants in newspaper before pressing: (a) if the plants are small, place many in the newspaper to make a full sheet; (b) if plants, leaves or flowering stems are long, bend them in a V-shape or N-shape; (c) if branches are large and bushy, trim the stem and some branchlets and leaves, taking care to leave general architecture intact; (d) hidden flowers, fruits or leaves can be exposed by removing some plant parts or bending leaves; (e) both sides of leaves and flowers should be displayed; (f) thick fruits and stems should be sliced into pieces no thicker than 2–3 cm.

thicker than 2–3 cm. Fruits, roots or stems which are thicker than this should be sliced and excess tissue removed to facilitate drying. This also permits the rest of the plant to be pressed flat. Long stems should be bent into a 'V' or an 'N' shape so that they can fit in the newspaper, whereas vines may be left in a 'U' shape to show their twining habit. Remember that when the plants are dried and mounted, only one surface will be visible. Therefore, arrange plants with both sides of the leaves exposed, some flowers opened and some fruits sliced in cross-section, longitudinally or in both ways.

Although these suggestions will allow you to arrange and press the majority of botanical species you find in the field, there are some plant families – such as the cacti, aroids and palms – which require special techniques. These families are often under-represented in herbaria, an indication that many field researchers do not have the necessary skills or time to prepare adequate voucher specimens. Yet these families often include plants of considerable cultural and economic significance. If you plan to make an extensive collection of these morphologically untypical specimens, first consult the literature or specialists for specific recommendations on how to press and dry the plants. A description of how to collect palms – some of the world's most economically important plants – is given in Box 2.4.

Box 2.4 How to collect a palm

With over 200 genera and 2500 species, the Palmae ranks as one of the most diverse plant families in the world. In many parts of Africa, Asia and South America, it is a family of great economic importance because it is a source of food, oil, fiber, construction materials and many other products [21].

Despite their ubiquity and cultural significance in the tropics, palms tend to be under-represented in herbaria throughout the world. For example, when Michael Balick and several of his colleagues looked for palm specimens in three herbaria of the Brazilian Amazon, they found that only 37.5% of local species were represented by collections [22]. This does not mean botanists and ethnobotanists are not interested in palms, but probably that they are intimidated at the thought of having to collect one. Palms are often large and spiny, have long leaves that do not fit neatly into sheets of newspaper and may produce massive inflorescences and fruits. How can we go about reducing them to standard voucher specimens?

In addition to following the general guidelines for plant collecting discussed in this chapter, palm specialists give several specific suggestions. First realize that you will probably need several sheets of newspaper for each duplicate. Because the leaf is often very large, three representative segments should be collected and each pressed in a separate piece of newsprint. The **apex** or upper portion of the leaf often contains fused leaflets and other distinctive characteristics. A middle section reveals the size, shape and distribution of the **pinnae** or subdivisions of the leaf. The **base** of the leaf, including its attachment to the stem, provides additional features that may be used to distinguish one species from another. You can collect similar representative portions of other important plant organs, including flowering and fruiting stalks.

In addition to calculating the height of the palm and the diameter of its stem, measure other distinctive parts that are not preserved whole in the herbarium specimen, such as the overall length of the leaf, inflorescence and fruiting branch. Whenever possible, take black and white or color photographs of the palm, including closeups of the leaves and flowering and fruiting stems as well as shots of the entire plant and its habitat. You may also count the total number of pinnae on a representative leaf.

Because palms are too big to be represented fully in an herbarium collection, extensive notes should be taken on any distinctive aspect of the habit, stem, sheaths, leaves, inflorescence or fruit that are not evident in the specimen. One important distinguishing feature of palms is whether they are solitary, consisting of one stem, or clustered, composed of multiple stems. The number, persistence and general arrangement of leaves may be distinctive. You should record these and other observations in your field notebook. Depending on the class of palms that you are documenting, you may wish to collect additional plant parts such as spiny stems, inflorescence bracts and sections of the trunk. Ideally, several fruits and flowers should be preserved in alcohol for later study.

John Dransfield, head of the palm section at the Herbarium of the Royal Botanic Gardens, Kew, is a specialist on the climbing palms known as **rattans** [23, 24]. Common in Southeast Asia, Africa and parts of South America, they provide the raw materials for a vast furniture-making industry and also have many non-commercial uses. As part of the Kinabalu Ethnobotany Project, John Dransfield visited Kinabalu Park in July of 1992 to train local collectors in the preparation of rattan voucher specimens.

In addition to the normal plant collecting equipment listed in Table 2.1, he recommends one additional element – plenty of time. For each rattan, the collector needs to prepare not only the portions of leaves, inflorescences and flowering stalks described above but also two lengths of stem for each duplicate – one containing the spines or other external covering and one that has been stripped to show the internal cane. If these stems have a diameter of less than 2–3 cm they can be dried in a plant press. If larger, they are placed on the screen above the gas stove. After a day in the dryer, the internal part of broader stems can be pushed out and thrown away. This decreases the drying time while leaving all important morphological features intact.

Dransfield also suggests collecting the distinctive spiny 'whips' which emerge from the tips of leaves and from leaf axils. These are normally collected as part of the leaf apex or base, but special attention should be given to obtaining well-developed characteristic whips for each duplicate.

Because they are armed with sharp spines, each whip is folded into a circle before pressing the specimen. As with all plants, samples of the flowering and fruiting stalks and additional amounts of fruits and flowers are gathered to complete the specimen. To avoid mixed collections, each separate plant segment is labeled with the collection number, written directly on the plant or on an attached jeweller's tag. A voucher specimen of a rattan collected during the Kinabalu course is depicted in Figure 2.5.

Figure 2.5 A pressed and dried specimen of *Calamus ornatus* Blume, prepared during a training course of the Kinabalu Ethnobotany Project. The four sheets contain a fruiting stalk with several loose fruits and the lower, middle and upper portions of the leaf. An entire spiny portion of stem and a stripped cane, dried outside the plant press, were integrated into the duplicate. Each portion of the plant carries the collection number, either written directly on the plant or on an attached jeweller's tag.

Almost all plants can be dried in a press. Nevertheless, some plant organs such as fleshy roots, succulent stems and fruits may be placed directly on the screen inside the dryer. Screen-dried plant parts should be later placed in paper bags marked in permanent ink with the same collection number as the corresponding pressed specimen. Alternatively, fleshy or over-sized plant organs can be placed in a spirit collection, further described below.

Make a sufficient number of duplicate specimens to fulfil your obligations to sponsoring herbaria, to specialists and to make a field reference collection, when desired. Normally five or more duplicates are pressed – one for a local reference collection, one to send to a specialist, one for a regional herbarium, one for a major national herbarium and one for a major international herbarium. Additional duplicates may be required if your work is sponsored by more than one herbarium or if you wish to have more than one specimen for the reference collection or for determination by specialists. Under each collection number, record in the field notebook the number of duplicates made for that species; this will later facilitate printing the correct number of labels, saving time and paper.

Once each set of plants is ready, the collection number is written on the newspaper, using an indelible ink pen or a wax crayon. The number is usually written in the lower right-hand corner, with the newspaper placed open side to the right. You may choose another location, but always write the number in the same place to make it easier to count specimens and put them in order. In projects where many collectors are involved, the initials of each collector should be placed in front of the number. For example, as shown in Table 2.2, José Rivera Reyes always precedes his collection number with his initials 'JRR' and Ricardo López Luna with 'RLL'.

2.1.6 Pressing the plants

Once the plants have been arranged and the number written on each newspaper, the collections are ready to be placed in the plant press. The press consists of three or four elements: (1) the press ends, which help keep the press rigid and square; (2) the straps, used to bind the press tightly; (3) ventilators, which allow the hot air to pass through the press; and, when possible, (4) felt blotters, which wick the moisture away from the plants and keep the ventilators from becoming damaged and flattened.

When necessary, a press can be made from local materials: the ends hammered together from strips of wood or cut out of wire screens, the ventilators dispensed with or cut from old cardboard boxes, the whole press held together by lengths of rope. When few materials are available, a simple press can be made by putting a small stack of extra newspapers between the pressed plants. These are placed between homemade press ends and are tied with whatever binding material is available locally. The small stacks of newspaper are changed every day, causing the pressed plants to dry slowly.

Materials for a higher-quality press can be purchased as funds become available. The best press ends are made from a hard wood such as oak, the individual laths held together by glue and rivets. The highest quality straps are similar to those used by parachutists. They are made of nylon and are fitted with quick-release buckles that allow the press to be adequately tightened and quickly opened. The most resistant ventilators are made from thick cardboard or from corrugated aluminum. When ordering or making ventilators, ensure that the channels run the width of the corrugates, because lengthwise channels will not allow hot air to pass through the press. Precut felt blotters of the appropriate thickness and other drying aids are also available from biological supply companies.

With these materials and the arranged plants at hand, the press is built. The straps are placed flat on the floor, with one press end on top. Next a ventilator (sometimes with a blotter) is added, followed by one duplicate. Another cardboard (and blotter) is added, followed by the next duplicate. When all duplicates of one set have been pressed, start with the next collection number. This process is repeated until all the day's collections have been placed in the press. The second press end is placed on top and the straps are brought around the press and tightened.

Experienced botanists offer many other tips for making a tidy press. As you are building the press, you will often notice that the pile starts to lean to one side or another. To correct this, you can reverse a section of plants, leveling out the pile. If there are thick plant parts in a certain collection, you may want to add an extra ventilator or two between plants, which helps to keep the plants flattened. Before the final tightening, you can straighten up the press by laying it on the floor on its various sides, pushing individual cardboards into line. At the same time, tuck in or remove any plant parts which are sticking out.

You not only see when a press is well prepared, but you hear it. Well-tightened straps have a certain 'twang' when they are plucked. That's the music that tells you it's time to get on with the drying.

2.1.7 Drying the plants

Some people dry the pressed specimens in the open air or in the sun, taking care every day to change the cardboards or stacks of newspapers that divide the plants. Because this is a tedious process that dries plants very slowly, most botanists prefer to place the press in a plant dryer. The dryer is a box, usually made of wood, that contains a heating element and a horizontal frame covered with metal screening. The frame is placed above the heat source to hold the press and to prevent plant parts from falling into the heating element and catching fire. The carpenter's plan for the plant dryers used in the Kinabalu Ethnobotany Project is shown in Figure 2.6.

The dryer works in a simple way. The filled plant presses are placed inside the dryer, resting on the screened wooden frame. Spare cardboards are placed over any part of the screen not covered by the press and crumpled newspaper is used to block smaller air passages. Unheated air enters through the ventilation holes at the bottom of the dryer, is warmed by the heating element and passes through the channels in the ventilators of the press, extracting the moisture of the plants. The hot moist air exits through the top of the dryer.

Many different heating elements are available. Choose one which is safe, efficient and easily maintained in your field site. A frame of six 100-watt light bulbs provides low-cost even heating but is dependent on having access to electrical current. Gas stoves that use butane or propane are an alternative, but the tanks must be refilled frequently. Kerosene heaters are commonly used, since replacement fuel is relatively easy to find in many parts of the world. Many other sources have been employed, including charcoal, solar heaters and microwave ovens.

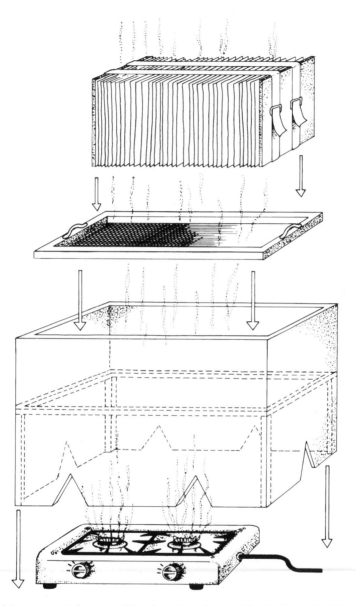

Figure 2.6 A carpenter's drawing of the plant dryer designed for the Kinabalu Ethnobotany Project. The ends and sides are made of half-inch plywood, whereas the inside frame is cut from 1 × 1 inch lengths of hardwood. The entire dryer is lined with glossy sheet metal and triangular ventilation holes are cut into the base. The metal screen and wood tray, placed inside the dryer, holds the plant press, prevents loose material from falling into the heating element and provides a platform for drying loose plant parts.

The dryer should be tailor-made for your particular needs. It should be wide enough for the press to fit snugly, high enough to leave sufficient room between the heating element and the screen and long enough to contain one or more full presses. Choice of the overall size and the construction material depend on how far you must travel with the apparatus, how long it will be used and how big are the plant presses. Ethnobotanists working on a short-term project that generates few collections can carry along a small portable dryer. If the project is based in a community, biological station or research institute, it is worthwhile making a sturdy, relatively permanent, large dryer that can be used by others after the initial project is finished. Several modifications increase the portability and efficiency of plant dryers. By using four pieces of thin fiberboard that are joined with corner hinges, you can make a lightweight dryer that is easily transported. Covering the inside with glossy sheet metal increases the intensity of the heat and the drying capacity.

Once you have built a dryer, test it to see if the plants dry well. If they turn brown or look burned, decrease the intensity of the heat or raise the frame further above the heating source. If the plants mold or look like they have been cooked, increase the ventilation in the dryer. This can be done by cutting triangular or semi-circular holes at the bottom of the dryer and by ensuring that the hot moist air can escape from the top. If the plants stay green and pliable for more than a day, then the heating element is not warm enough or there is too much ventilation, allowing the hot air to escape before it dries the plants.

Sometimes the problem is with the plant press and not the dryer. Check the condition of the cardboards from time to time. The channels eventually become crushed and the hot air cannot pass through, causing the plants to cook or to rot.

How long should the plants be left in the dryer? The answer varies according to the quality of the dryer and the press as well as the succulence of the plants collected. With good equipment, many herbs and grasses will dry within 24 hours. Thick leaves, fleshy fruits and many-petaled flowers will take more time. Drying a cactus, orchid or other succulent plants can take several days. Special techniques, such as carving out fleshy tissue from inside the stems, are used to decrease the drying time of these specimens.

Most botanists turn the plant press in the dryer every 12 hours so that both sides of the plants are evenly dried. They open the press and review the plants every day, removing the specimens that are dry while leaving the still humid ones for additional drying. They use a simple test to see if the plant is fully dried. When a leaf or flower is bent over, it should snap. If still pliable, it needs additional drying. Fruits should be hard to the touch and not at all mushy. With a little experience, collectors can see if a plant is dry just by looking at it and by occasionally testing a fruit or an especially thick flower or leaf.

Once plants are removed from the press, all duplicates belonging to a numbered set should be assembled in a single bundle, which is then wrapped with an

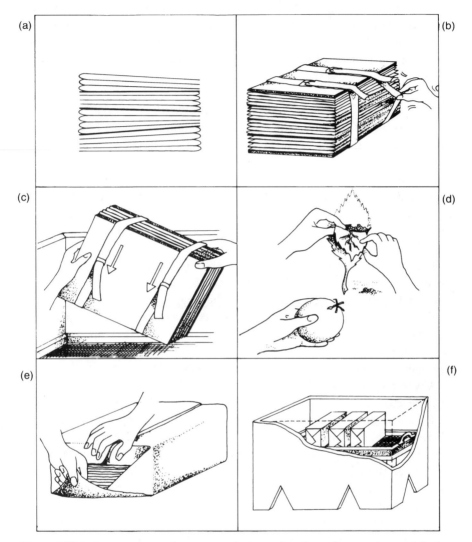

Figure 2.7 Various suggestions for pressing, drying and packing plant specimens: (a) when building the press, correct for any tilting of the stack by turning some specimens around; (b) pluck the straps of the press to hear if they are sufficiently tightened; (c) turn the plant press in the dryer twice a day, tightening the straps as the plants dry; (d) check for dryness by bending a piece of leaf, which should snap, and by pressing on fruits or fruit sections, which should be hard; (e) when plants are removed from the press, place all duplicates together in a single packet and then wrap them snugly in doubled whole sheets of newsprint to create packages of 5–15 collections; (f) if packages have been exposed to humidity and insects, place them sideways in the dryer for 24 hours before packing them in plastic bags and cardboard boxes.

additional half sheet of newspaper that prevents material from falling out. Any fruits or other plant organs which have been dried separately are placed with the duplicates. The bundles should be protected from insects and moisture by placing them in sealed plastic bags. In humid areas, the dried plants can be left in a covered box on top of the dryer, which gently warms them and keeps them from becoming damp.

2.1.8 Labeling the specimens

After an extended collecting trip or period of fieldwork, you are faced with the task of labeling the collections. The label contains most of the information recorded in the field notebook for a particular specimen. When the dried plant is glued to mounting paper in the herbarium, the label is attached in the lower right-hand corner. It provides other researchers with the basic information that they wish to know about the plant – its scientific name, where and when it was collected, what it looked like in the field and the name and collection number of the collector.

The most efficient way to turn data from a field notebook into labels is by creating a computerized database that is formatted to print the information just as you would like to see it on the specimen. The database has the added advantage of allowing you to update identifications of the collections and search for particular species or localities very rapidly. If you do not have access to a computer, you will have to prepare each label on a typewriter or by hand, which is a tedious operation especially when many collections are being processed.

Any terms abbreviated in the notebook should be written in full on the label. The collector's name should be completely spelled out. Abbreviations for middle initials are acceptable, but first and last names should be given fully. In countries where two last names are used commonly, such as in much of Latin America, both should be spelled out. Consider, for example, three fictional Mexican collectors: José López Ruiz, Juan López Rodríguez and Josefina López Ramos. If these collectors' names are written or cited as J. López R., there will be confusion about who collected the plant.

The date should be written out fully in an unambiguous way. In many countries, '6/5/92' means 'the 6th of May, 1992' but in the United States it means 'the 5th of June, 1992'. Because the collection of herbarium specimens has been common practice for over two centuries, it is best to specify the year fully: '92' could be 1792, 1892 or 1992. When entering these data in a computerized data base, some researchers give the date as year/month/day. For example, 'the 6th of May, 1992' would be written as 92/5/6. This simplifies sorting of the collections according to date.

Whether the labels are prepared using a computer printer or a typewriter, you must ensure that they are of archival quality, which means that they will last indefinitely. This requires that high-quality acid-free paper and permanent ink be used. Acid-free paper can be purchased in standard-sized sheets, which allows six to eight labels to be printed on each sheet. Most inked cloth ribbons provide stable print, but some laser printers, photocopiers and tape ribbons give print which fades or rubs off over time.

In addition to the collection label, you or other researchers may attach small annotation labels to the specimens to correct the scientific name of incompletely or improperly named specimens. Annotation labels are also attached to indicate that: (1) some material has been removed for phytochemical or anatomical analysis; (2) the specimen is a voucher for ecological or ethnobotanical research; or (3) additional material is stored in boxes or in a spirit collection.

It is important to indicate that the specimen is an ethnobotanical voucher, but opinion varies on whether or not to give detailed ethnobotanical information on the label. Some researchers (who point out that ethnobotanists sometimes refer to herbarium labels in their search for information on local names and uses of plants) include a full print-out of all the information contained in their ethnobotanical database. Others prefer to include a small annotation label which refers to a publication that contains detailed ethnobotanical information. They argue it is necessary to consult a publication in order to be able to understand fully the cultural context of the use as well as the orthography to write the local name.

Once the labels are printed and cut to size, they are inserted in the collections, one per sheet. As you place the labels with the plants, cross-check the information, giving special attention to verifying if the collection number and scientific name match the specimen. This is the best opportunity to detect an error in the numbering system that would be difficult for an herbarium technician to discover. In addition, make sure that each collection is complete. If you find one duplicate that is sparse or sterile, you may be able to transfer flowers, fruit and foliage from another collection that belongs to the same numbered set.

2.1.9 Distributing the specimens for identification

While placing the labels, you can sort the duplicates into piles for the various institutions and applications – the first set for an institution in the host country and another set for your institution, some specimens for determination by special-ists and others for a reference collection to be used in further fieldwork.

As you sort the specimens, keep a list of the collection number of all specimens sent to each institution. Updating identifications and telling colleagues where to find taxa of special interest to them is easier when you know where you deposited the specimens. In addition, many scientific journals require that ethnobotanical information be documented by mentioning the collector, number and location of any specimens referred to in the text.

Once the duplicates have been sorted into piles, you should count how many specimens will be sent to each institution. When dealing with large numbers of collections, an accurate count can be made by sorting the piles into groups of ten.

The counted stacks can then be tightly wrapped in full sheets of newspaper, which are taped shut. The number of specimens and the acronym of the institution to which they will be sent is written on the packet.

The tightly wrapped bundles of plants can be placed sideways in a plant dryer for 24 hours. This dries out any ambient humidity and reduces the possibility of insect damage during storage and shipping. Upon removal from the dryer, wrap the collections in plastic bags and place them snugly in sturdy boxes. Any extra space should be filled with packing material, such as crumpled newspaper.

Before sealing each box with binding tape, insert a shipping notice that contains your name and address as well as the address of the receiving institution. The destination and return address should also be written clearly on the outside of the box. The shipping notice should also list the number of specimens being sent, the region in which they were collected, the total number of boxes shipped and which other herbaria are receiving duplicates of the same collection. Indicate that the collections are ethnobotanical vouchers and note any special treatment the plants have received, such as if they were treated with alcohol before being dried.

2.1.10 Making a spirit collection

The best way to make an ethnobotanical voucher is to press and dry a fertile plant specimen. But in some cases, you will find that this is not possible. When making an inventory of food plants sold in a marketplace, rarely will you encounter a whole fertile plant. Instead, you will often collect fruits, nuts, roots, leafy shoots and flowers. While some of these plant organs could be pressed and others dried in paper bags, another approach is to bottle them in a liquid preservative. Many researchers simply use 90–100% alcohol, but more elaborate mixtures are available. Some of these are described in *The Herbarium Handbook*, which explains techniques of plant collection and herbarium curation used by botanists of the Royal Botanic Gardens in Kew, England [25]:

> The preservative used at Kew is a mixture of 53% industrial methylated spirit (i.e. ethanol + 9% water + 2–4% methanol), 37% water, 5% formalin (dilute formaldehyde) solution in water and 5% glycerol. This may become warm while being mixed, but is quite safe. Alternative mixtures can be used such as 70% ethanol, 29% water and 1% glycerol or FAA (formaldehyde, industrial methylated spirit and acetic acid), which however, makes specimens brittle. Particular care needs to be taken with alcohol, which is highly flammable. Formaldehyde is toxic and careful handling is required. Hands must always be washed after contact with it and it must be used in a well ventilated place to prevent inhalation of its fumes. It is advisable to wear goggles.

The plant material is placed in a jar and the preservative poured in until all parts are covered. A label written in permanent ink is placed in the liquid or is glued to the outside of the jar. It should contain all of the information included on a regular herbarium specimen. The jar is covered with a tight-fitting plastic lid and the level of the preservative checked and topped up every few months.

Spirit collections are bulky, heavy, somewhat toxic and rather expensive to prepare. Yet the trouble is often worth the advantages. Pickled fruits and flowers are used by taxonomists to write accurate botanical descriptions and by plant illustrators to make realistic drawings. The collections may be used as props in ethnobotanical fieldwork. Although the plant organs lose some color, they maintain their shape, making it easier for local participants to identify them.

Not all herbaria have facilities for maintaining a spirit collection. Before sending pickled material to an herbarium, ask the curator if he or she is interested in curating the collections.

2.1.11 Collecting wood and bark samples

Wood samples are collected to verify the identity of the species used to make a particular artifact, to document the wood anatomy of poorly known trees or to conduct stress and drying studies of species that have potential value as timber. Collections of bark add to the value of a voucher specimen, often facilitating its identification.

The study and identification of wood requires skill and experience. Samples should be sent to a qualified anatomist who has agreed to receive the material and has provided special instructions on collecting and preserving the sample. To document the structure and quality of a wood which has not been previously studied, you should seek out the tree in the field, make a fertile voucher specimen and collect a chunk of mature wood, accompanied by pieces of the bark. It is preferable that samples include some heartwood which is already many years old. If no felled trees are available, cut a wedge out of the trunk, as shown in Figure 2.8. If this is not possible, collect a section of a branch, ensuring it is at least several years old.

If you need a sample of wood from an artifact, take care to limit damage to its function or beauty. When possible, the wood should be taken from another source, such as a damaged or discarded item made from the same kind of material. Alternatively, fresh material could be collected in the field with someone who makes the artifact. Going to the field with a local expert will yield much more information about how the wood is handled – when and where the trees are cut, how the wood is dried, how the object is made and other details of the process. It will also allow you to collect a voucher specimen of the tree, facilitating its identification.

Ben J.H. ter Welle, a wood anatomist from the Netherlands, provides a succinct description on how to collect bark and wood samples [26]:

Making bark samples is very simple. A sample of approximately 1.5 × 1.5 × 1.5 cm is sufficient, unless the bark is very thick. A sample should include a

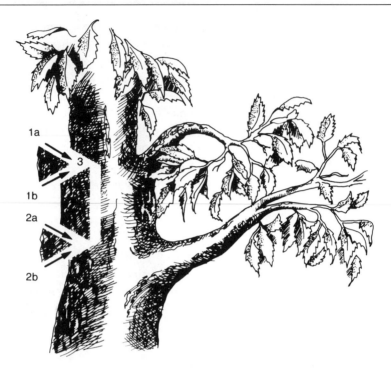

Figure 2.8 Ben J.H. ter Welle's illustration of how to cut a sample of wood from a tree trunk or branch. Triangular cuts, indicated by the numbers 1a, 1b, 2a and 2b, are made above and below the piece of wood that is to be collected. Cut number 3 frees the piece of wood from the trunk. In most cases this method of wood collection does not kill the tree.

small piece of wood, as the cambial area might provide some extra information. The samples should be collected in 70% alcohol or FAA.

Wood samples should contain adult wood.... In general a wood sample from trees of about 12 cm long × 8cm wide × 4 cm thick is sufficient. For treelets, shrubs and herbs with a woody base, a sample of about 12 cm long from the main trunk is sufficient... for lianas the same length of sample is needed. If possible, both a sample from the trunk near ground level and one from the canopy should be collected.

Wood samples can be preserved without any specific treatment. Solar heat is all right as well as various types of stoves. Do not dry them too fast. Fungal attack is not a problem at all, since the anatomy of the wood is hardly affected. Fungi occur only when the sample is wet or green.

2.1.12 Making living collections

Making voucher specimens can involve killing plants. Keeping a plant alive requires a distinct approach and serves different purposes. Living material is referred to as **plant germplasm**, defined as all the genetic material in a plant. In practice, most researchers use the term to refer to the different plant organs – seeds, stem cuttings, tubers, bits of leaf tissue – that can be used to propagate a botanical species.

Germplasm may be collected during an ethnobotanical project to: (1) enrich the collections of a botanical garden; (2) contribute to the conservation of an endangered population of plants; (3) supply material for horticultural or agricultural experimentation leading to cultivation of the plant on a wider scale; (4) ensure the preservation of the genetic diversity of crops, their relatives and other economically important plants; or (5) provide stock for distribution to other communities that wish to cultivate the species.

Each of these goals requires well-planned coordination with colleagues in germplasm centers, botanical gardens, agricultural experiment stations or other institutions. When collecting voucher specimens, you can afford to store material for months before sending it to collaborating institutions. With living material, you do not have this luxury. Unless cared for properly, seeds lose their viability, fleshy organs rot and cuttings dry out. Before starting a living collection, work out the logistics of sending the material to the receiving institution and consult your colleagues for precise information on how each species should be sampled and documented. Be aware that special permits may be required for collecting live plants and transporting them across international borders.

Providing live collections for cultivation in a botanical garden for purely demonstration purposes is straightforward. You simply gather seeds, cuttings or whole plants from a representative population and deliver them to the person responsible for accessions. Voucher specimens should be made to facilitate identification of the species and to serve as a permanent record of the collection.

Collecting living material for other purposes requires careful attention to the genetic diversity of a species. Most researchers who collect germplasm focus on a particular crop and its wild relatives and they travel over a vast area – often the plant's center of origin and diversity – to sample randomly as many local populations as possible. Conservationists visit populations throughout the range of a species they are seeking to protect.

Ethnobotanists and local people who are working together to make an inventory of useful plants rarely have the opportunity to make a broad collection of germplasm or to visit the entire geographical range of a species. Yet they have much to contribute to studies of genetic variability of plants, because they can provide data and material unavailable to colleagues who visit a community for just a few hours or days. While germplasm expeditions rarely venture far from major routes or explore thoroughly a single region, ethnobotanists and their local counterparts have the ability to visit remote locations and find previously uncollected populations, guaranteeing a complete sampling of the local varieties and wild relatives of crops and other plants of interest. As part of the ethnobotanical inventory, the environmental parameters and agricultural practices associated with each plant population are documented in detail. Collections are made throughout the year, ensuring that seeds, tubers and other organs are collected at the peak of their maturity. If the inventory continues for a number of years, collections of particular interest can be replicated.

Ethnobotanists have much to gain from making good quality germplasm collections and sending them to specialists. In local ethnobotanical classifications, most crop plants are over-differentiated – they are divided into many varietals that correspond collectively to a single scientific species. Plant taxonomists can provide only a species level identification of these varieties, whereas agronomists who have a specialized knowledge of the crop usually classify them in a finer way. They can provide valuable information about the origin, geographic distribution and agricultural practices associated with each variety.

The exact method of sampling should be discussed with representatives of the institution that will curate the live material [27]. As a minimum, you should collect material from 10 to 50 randomly selected plants in each population sample. For a complete collection of germplasm, several such samples should be taken for each species. 'Population sample' must be defined for each species, taking into account various factors:

- The **folk varietals**. Traditional agriculturalists name the local variants of crops and other cultivated plants, sometimes including wild relatives. Make a collection of each folk variety, noting how it is different from other cultivars of the same crop.

- The **ecological zones** of the community. The same varietal may grow in several different locations and climatic zones recognized by local people. For example, the Chinantec have a variety of white corn that grows in hot dry ranches, in the dry temperate zone, along cleared slopes in the humid cloud forest and in the hot humid lowlands. Each of these variants would be sampled, even if they carry the same local name.

- The local **agricultural practices**. Among crop plants, there are some cultivars which are planted in tilled fields and others in newly cleared areas. Some traditional varieties may be maintained in home gardens, whereas others are planted in polycultures or as pure stands in cultivated fields. After discussing the range of agronomic practices with local people, gather samples of each varietal grown in different ways.

- The **cultural diversity** of the local population. Even when the community is composed of a single ethnic group, there are apt to be differences between the plant varieties cultivated by different social groups. Some of these differences may simply result from the distinct ecological zones in which they farm or the agricultural practices they employ. Other differences could be linked to the sharing of seeds or tubers among members of the same kin group or the passing down of cultivars from one generation to another.

2.2 Preparing an ethnobotanical reference collection

Many ethnobotanists keep one duplicate of each specimen which they mount and use as a field reference collection for eliciting more data on the classification, use and preparation of the species [28]. The typical herbarium sheet, approximately 30 cm wide by 45 cm long, is too large for use in most field projects. An easy to handle specimen that is representative of the species can usually fit on mounting paper that has been cut to normal notebook size, approximately 20 × 25 cm. Two pieces of this size can be cut from a single sheet of mounting paper.

Adequate mounting of specimens takes time and attention to detail. The dried, pressed plant is first fitted on the sheet, ensuring that all parts fit within the borders of the paper. White glue is then placed on the side of the plant which will be attached to the paper. After all the specimens are glued, they are placed between cardboards. Pressure is applied overnight, either by gently closing the specimens in a press or weighing them down with heavy books. Cloth tape can be used to secure leaf and flower stems or these can be sewn down with needle and cotton thread. The finished specimens are protected by placing them in plastic sleeves that can be arranged in a ring binder. A black and white or color photograph may be taken of the specimens to provide a permanent record of the appearance of each specimen, which may be of interest to other researchers.

The rationale behind the use of a reference collection is that a broad range of plants can be shown to many different informants over a short period of time and each specimen can be presented with flowers, fruits, leaves and other characteristic plant parts. In nature, only a relatively small portion of the flora is fertile at any one time and it is time-consuming to walk through a forest one-by-one with a large group of respondents. A disadvantage of using the reference collection is that informants cannot smell the leaves, taste the fruits, see the overall habit of the plant or sense many other aspects that are evident when it is perceived in its ecological or cultural context. These disadvantages can be ameliorated somewhat by adding props, such as dried fruits, pieces of wood or a drawing showing the aspect of the whole plant or a close-up photograph showing a particular feature which is characteristic of the species.

After the reference collection has been used to elicit ethnobotanical data, it can be left in the community as an educational tool. For this reason, use archival materials in the preparation of the specimens, including acid-free mounting paper and labels, good quality glue, linen tape and polyethylene envelopes. Avoid polyvinyl chloride envelopes and standard paper, which slowly release acid, particularly when they come into contact with water.

2.3 Herbaria and the curation of plant specimens

Herbaria are institutions where botanical specimens are kept. The botanists and technicians who work in herbaria perform a variety of tasks that range from keeping the collections free from insects and moisture to filing them in such a way that facilitates retrieval. This ensures that the collections will last for a long time – up to hundreds of years – and that they can be consulted by people who are carrying out studies on a particular taxonomic group or on the flora of a specific region [29].

Once the labeled plants are received, the **curation** – or care of the specimens – begins. The plants are fumigated to kill insect pests. They are then glued onto acid-free mounting paper to which the label is attached. If not completely identified, the specimens are given to a **generalist**, who can recognize a broad range of plants by sight and who may decide to send certain plants to **specialists**, experts in particular taxonomic groups. After being fully identified, the specimens are filed in a cabinet where they are protected from insects, dust, sunlight and other damaging agents. Afterwards, the plants usually leave the herbarium only if they are sent to a specialist working at another institution.

Major herbaria contain many hundreds of thousands of collections. The botanists working at these institutions are often interested in collaborating on ethnobotanical projects, especially if they will receive plant specimens from a taxonomic group or region of the world that interests them. This sponsorship may take a variety of forms. The botanists may come to the field to collect with you or they may agree to accept a duplicate set of plants for identification. On occasion, they can provide collecting materials, financial support or advanced taxonomic training for some project personnel.

Collaboration should be arranged well in advance of the fieldwork. It is best to contact specialists well before sorting and shipping specimens to enquire if they would like to receive material in the taxa that they study. Some taxonomists receive hundreds or thousands of specimens for identification each year, which means that you will probably have to wait months or sometimes years before receiving a list of the correct scientific names for your collections. The list that you are eventually sent will give the collector's name and number followed by the complete scientific name of the specimen. This information should be recorded immediately in your field notebook or in your computerized database.

When collecting voucher specimens, remember that herbaria are set up to curate plants that fit on standard herbarium sheets. You should send only plants that are pressed in newspaper, labeled and carefully wrapped to prevent damage during shipping. Large fruits and other over-sized plant parts that are kept in boxes or in a spirit collection are also accepted by most herbaria, particularly if they are cross-referenced to a specimen that will be mounted.

Collections without labels or plants which are sterile, moldy or otherwise poorly preserved are given low priority in an herbarium. Some botanists will agree to identify such specimens but will not curate them, which means they will not be mounted and filed in the herbarium. If not thrown away immediately, they will be relegated to storage. In either case, they will not be accessible as ethnobotanical voucher specimens.

Herbaria do not typically accept remains from archaeological digs, fragmentary material found in markets, agricultural seeds and tubers or artifacts made from plants. There are specialized institutions that will accept and identify these materials. **Seed banks** accept material which can be propagated, usually of crop plants. The personnel are often willing to provide identifications at a subspecific level and may be interested in participating in agronomic surveys. **Ethnobotanical museums** and **gardens** accept living or preserved collections, including artifacts made from plants. **Archaeological museums** will curate and identify plant remains, but these must be taken from an archaeological site using specialized techniques. As with herbaria, it is essential to make early contact with personnel from these institutions to discuss the conditions under which they are willing to accept donations of plant specimens and identify them.

2.4 Judging the completeness of a plant survey

When collecting the useful plants in a community or region, you may ask yourself occasionally how complete is your inventory [30]. How many folk botanical categories have been documented and what portion of the local flora has been collected? How long do you need to go on collecting, hoping to find a new name, use or species not yet recorded? No study can be exhaustive, because this would require that we ask informants about every plant or animal in the community, a task that is difficult in temperate zones and nearly impossible in tropical areas. Yet even when seeking to collect only a narrow range of plants, such as the trees and shrubs used for firewood or the medicinal plants used to reduce fever, we should demonstrate the diversity of botanical and folk taxa represented in our collections.

The extent of coverage for each folk or scientific rank can be judged in three dimensions: breadth, depth and replication. **Breadth** is the percentage of the total number of categories at a given rank – botanical family, folk lifeform and so on – which have been collected. **Depth** is the percentage of the total number of subordinate categories of that rank which have been sampled. **Replication** is the number of times that each category has been documented. In Table 2.3, I

Table 2.3 The measurement of breadth, depth and replication of botanical collections at the rank of botanical family

Breadth	Depth	Replication
Number of families collected divided by total number of families in the region	Number of genera collected in each family divided by total number of genera in that family; summation and division by total number of families gives a cumulative measure	Total number of collections per family; summation and division by total number of families gives a cumulative measure

summarize how these concepts can be employed at the rank of botanical family. This example can be extended to other scientific and folk botanical ranks as well as to use categories.

Although these measurements are laborious to calculate for an entire flora, they can be made quickly for a limited number of categories. By taking the time to estimate the breadth, depth and replication of collections in terms of both scientific and folk categories, we can guide our collecting activities, visiting certain ecological zones that are undersampled and looking for specific folk categories or uses for which voucher collections do not yet exist.

As discussed in Chapter 1, we can take both a static and dynamic perspective when analyzing data. The static approach is equivalent to taking a snapshot of the results at a certain point in the study. We calculate the proportion of folk or scientific categories that have been collected and present the results in a table. Another technique is needed if we are to measure the efficiency of our collecting, that is, the rate at which new categories are detected as the inventory continues. This perspective is gained by judging how the proportion of total categories sampled increases as a function of the cumulative number of collections made. The data are plotted onto an xy graph and the resulting curve shows the rate at which new information is being added.

In order to calculate the completeness of ethnobotanical collections, we need first to estimate local botanical diversity, measured in terms of the number of families, genera and species in the region. In addition, we need to gauge the numbers of folk categories in the local classification, particularly at the lifeform, generic and specific ranks. Researchers refer to the full range of scientific categories as the **flora** and to the set of folk categories as the **ethnoflora**.

2.4.1 Assessing the breadth of a collection

When looking at the breadth of a collection, it is best to start with the static approach. After estimating the diversity of the local flora and ethnofloras, you simply count how many scientific and folk categories are represented in your voucher collection. The breadth is then calculated by dividing the number of categories you have detected at a certain rank by the total number of categories at that rank. Let's say that there are 150 botanical families in the region where you are working and the 500 ethnobotanical specimens that you have collected correspond to 75 of them. The breadth of your collection, in terms of botanical family, is 75/150, which equals 0.5. That is, 50% of local botanical families are represented in your voucher specimens.

After taking these 'snapshots' of the breadth of the collections, you can judge at what rate new categories are being added to the inventory. This calculation is more tedious, because all the collections must be considered in order, from the first to the last collected. A taxon is counted only the first time that it appears in the collection and its position is noted.

Table 2.4 The appearance of new families in the first 30 collections of a fictional set of 500 ethnobotanical vouchers; * marks the first appearance of a family

No.	Family	No.	Family	No.	Family
01	*Asteraceae	11	*Labiatae	21	Rubiaceae
02	*Fabaceae	12	*Scrophulariaceae	22	*Lauraceae
03	*Solanaceae	13	Asteraceae	23	Asteraceae
04	*Rosaceae	14	*Orchidaceae	24	Poaceae
05	*Rubiaceae	15	Solanaceae	25	Solanaceae
06	*Ericaceae	16	Fabaceae	26	*Melastomataceae
07	Asteraceae	17	*Onagraceae	27	Labiatae
08	*Poaceae	18	Rosaceae	28	*Euphorbiaceae
09	*Piperaceae	19	Ericaceae	29	Fabaceae
10	Fabaceae	20	Labiatae	30	Scrophulariaceae

Let's say that you are calculating the rate of appearance of new families in the 500 ethnobotanical vouchers mentioned above. The first 30 collections are shown in Table 2.4. In the first set of ten collections, eight new families are detected; in the second set of ten there are four new families; in the third set, three new families. This calculation is continued until all collections have been reviewed and all new appearances of family recorded.

Counting the first time that scientific taxa appear is straightforward, but calculating the same for folk categories is more complex. We must define which indigenous plant names correspond to true folk taxa and which are merely synonyms or descriptive phrases, a process that is particularly time-consuming for categories at the specific and varietal ranks. After verifying the folk identification of each specimen, the counting of new appearances of folk categories proceeds as outlined above for scientific categories.

2.4.2 Assessing the depth of a collection

Although breadth is a good measure of the thoroughness of an ethnobotanical collection, you may wish to make a more detailed estimate by seeing how well the internal diversity of each taxon is represented in your voucher specimens. You might ask, for example, what portion of genera was sampled in each of the 75 botanical families detected in the set of 500 collections mentioned above. If each family is represented by collections of just one genus, you would have a shallow understanding of how these families are classified and used by local people. You would not be in a good position to discuss, for instance, how the Asteraceae is classified or the diverse uses of the Labiatae. Ideally, every genus in the flora would be collected at least once, but more likely you will find that the proportion of genera sampled varies for each family.

A similar analysis can be carried out for folk categories. You may have collected 400 folk generics, but do you have vouchers that correspond to each of the folk specifics in each generic? You have collections of corn, beans and squash, but have you documented each named folk varietal with a voucher?

The calculation of depth is simple, if time-consuming. For each taxon at a given rank, the number of subordinate categories sampled is divided by the total number of subordinate categories found in that taxon. For example, if there are 25 genera of Asteraceae in the local flora and five are represented in your voucher collection, the depth is $5/25 = 0.2$; that is 20% of the Asteraceae genera have been sampled. For a cumulative measure, all depth values are added together and the sum is divided by the total number of taxa.

2.4.3 Assessing the degree of replication in a collection

Science is partly based upon the ability to repeat each observation and experimental step and to see if the result is always the same. In ethnobotany, we confirm our results not by repeating experiments, but by replicating collections – preparing voucher specimens of the same folk or scientific taxon on several different occasions, with different local people. Because there may be thousands of taxa to sample, replication is always a matter of degree. Many categories are documented by one or a few collections and only a few categories are represented by many collections.

When analyzing a set of ethnobotanical collections, we can prepare a bar graph that shows how many specimens have been collected for each category at a given rank, as explained in Chapter 1. If the collectors have widely sampled the local flora, we will invariably discover a characteristic long-tailed curve, indicating that a few taxa have been collected many times, while the majority of taxa are represented by just a few collections. In an extended collecting program, we find that this graph maintains its long-tailed shape, but that the curve is progressively displaced to the right as the number of replications per category increases. If we instruct the collectors to concentrate on the taxa which have been least collected over the course of the project, we find that the slope of the curve gradually becomes flatter, reflecting that a similar number of collections have been made for each taxon.

Charting the degree of replication in a set of collections helps us to assess how much confidence we can have in our data. The more times that we collect the plants that correspond to a folk category, the more certain we are that we understand how the taxon is delimited. Similarly, the more specimens that we have of a botanical family, genus or species, the more sure we are of how the local people classify these taxa.

2.4.4 How many collections are necessary?

The measurements illustrated above give us a great deal of information about a set of ethnobotanical vouchers which has been systematically collected. Analyzed together, the measurements allow us to assess just how much we need to collect

Table 2.5 The estimated number of ethnobotanical collections required for various levels of documentation of local knowledge in one community of the Sierra Norte flora

Number of collections	Type of analysis	Level of documentation of ethnobiological ranks
> 750	Outline	All lifeforms; many vouchers of generics, specifics and varietals
>1500	Sketch	All lifeforms and generics; additional vouchers of specifics and varietals
>2500	Detailed study	Prototypical species of folk botanical categories of all ethnobiological ranks
>5000	Definitive study	Prototypical species and extended ranges of all folk categories
>10 000	Complete analysis	Documentation of all local plant species, including those not classified or used by indigenous people

before feeling confident that the local flora and folk categories have been adequately documented. Table 2.5 indicates the range of collections necessary for the various stages of an ethnofloristic inventory in communities of the Sierra Norte of Oaxaca, given local ethnic and floristic diversity.

Additional information on ethnobiological ranks is given in Chapter 7. This assessment could be modified for other regions of the world, increasing or decreasing the levels according to the species richness of the flora, the complexity of indigenous knowledge of plants, the number of ethnic communities and other factors.

3

Ethnopharmacology
and related fields

Figure 3.1 Benedict Busin, local coordinator of the Kinabalu Ethnobotany Project, testing a plant extract for presence of alkaloids. With a simple set of reagents and glassware, screening tests can be carried out in the field, contributing to the discovery of plants with substances showing interesting biological activity.

3.1 Proceeding with a phytochemical analysis

3.1.1 Getting involved in phytochemistry

Laboratory analysis of medicinal and other useful plants is a costly and time-con-suming endeavor. Because there are many different chemical compounds, botan-ical species and potential uses, colleagues who study the chemical components of plants must decide which species are apt to yield the most promising results before they proceed with extensive investigations.

If you are interested in taking a phytochemical approach in your fieldwork, first refer to the chemical and ethnopharmacological literature to discover existing knowledge about the various species in your ethnobotanical inventory. You will be wasting time and money if you blindly duplicate tests which have already been carried out by other researchers. A good place to start is with *Trease and Evans' Pharmacognosy*, a standard reference which gives many examples of well-known drug plants [31]. A simple description of the diversity of compounds found in plants is given in Box 3.1.

Box 3.1 Compounds found in plants

In order to grow and reproduce, plants need relatively large amounts of certain elements, called **macronutrients**. Some of these are elements which are derived from air and water, such as hydrogen, carbon and oxygen. Other essential elements, including potassium, phosphorus and nitrogen, are pri-marily absorbed from the soil. Macronutrients also include complex mole-cules such as amino acids, sugars and carbohydrates that are made by living organisms.

Trace elements, which are needed in small quantities for plant growth, are called **micronutrients**. These include some heavy metals such as copper, manganese, zinc, iron, cobalt and other substances taken from the soil. When plants die and decompose, they release both macronutrients and micronutrients into the environment. Researchers who study agroecosystems are interested in determining the relative quantity of these nutrients in different species of plants, particularly those which are used as natural fertilizers.

Some natural compounds, called **primary metabolites**, are produced by almost all plants because they are essential to the biochemical pathways that control growth, photosynthesis, respiration, flowering and other basic processes. These substances include carbohydrates, proteins, fats and nucleic acids, which are the some of the basic components of human nutrition. **Plant physiologists** study the role of these primary metabolites and other substances – including hormones and enzymes – in how a plant functions. Nucleic acids, called DNA and RNA, carry the genetic informa-tion that forms the master plan behind all plant processes. **Plant geneticists**

---

investigate how plants are regulated by this basic genetic material and the proteins it produces.

Although primary metabolites are relatively uniform in structure and are present in almost all plants, **secondary metabolites** – often responsible for the characteristic smells, colors, flavors and medicinal properties of plants – are remarkably diverse and are distributed throughout the plant kingdom in characteristic patterns. Although some may simply be the waste products of physiological processes, most of these compounds aid plants in adapting to environmental conditions, competing with other plants, and either warding off attacks by predatory insects and animals or attracting ones that play a role in pollination, fruit dispersal or protection. **Essential oils**, for example, help reduce water loss in plants growing in arid zones, repel insects and deter grazing animals. Some **alkaloids**, bitter-tasting compounds that are often poisonous, discourage predators. For plants that grow in poor soils and cannot recycle nitrogen or derive it from bacteria, alkaloids serve as a storehouse of nitrogen. These and many other secondary compounds are the main focus of natural products chemists, pharmacognosists and ethnopharmacologists.

Comparative phytochemistry, the study of the distribution of both primary and secondary metabolites in the plant world, guides the search for not only new plant compounds but also new sources of known drugs. It is central to the work of some botanical systematists, called **chemotaxonomists**, who seek to clarify the relationship between different botanical taxa of various ranks by looking for the presence or absence of certain phytochemicals. For example, taxonomists were uncertain in which botanical order to place the cactus family until chemotaxonomists discovered that cacti contain betalins, a compound which is characteristic of plant families in the Centrospermae. Many similar examples have allowed systematists to confirm or modify the current classification of plants.

The search for information is greatly facilitated if you have access to major computerized databases on plant chemistry, pharmacology and medicinal plants that are housed at some research institutions. For example, the NAPRALERT database developed at the University of Illinois at Chicago contains information on the chemical constituents and pharmacology of biological organisms, drawn from more than 115 000 articles and books. The database includes the results of studies on more than 34 000 species of flowering plants, 2200 types of fungi, 2300 kinds of animals and many other living organisms. This information has helped guide ethnopharmacological research in many countries, particularly in Latin America.

After this search for information and consultation with colleagues who work in laboratories, you can begin fieldwork by carrying out some initial screening in the field and by collecting samples which can be further tested at research institutes. The first step in the process is to select, through an appropriate **sampling method**, the plants that will be analyzed. Small amounts of bark, leaves, roots, fruits or other parts of the plants are **screened** to see if they show activity in specific biological or pharmacological test systems or if they contain certain compounds which tend to be physiologically active or economically valuable. If a plant shows promise, you can proceed to make **bulk collections**, gathering several kilograms of material that is brought to the laboratory for detailed analysis.

If you have a particular interest in phytochemistry, you can spend some months in a laboratory, learning basic techniques and gaining familiarity with specialized equipment. Carrying out the entire procedure of phytochemical and pharmacological testing requires advanced training and access to sophisticated laboratory equipment. Whether you are just sending plant material for analysis or are planning to do some analytical work yourself, be aware that each laboratory is set up to run different bioassays. Because both private and public laboratories may seek to obtain patents on unique compounds that may become profitable, they do not usually disclose the specific bioassays employed.

3.1.2 Colleagues in laboratories

If you wish to begin a study of the chemical constituents of useful plants, you will find yourself collaborating with colleagues from a diverse range of academic fields such as pharmacognosy, ethnopharmacology and natural products chemistry. What are the basic differences between these fields?

Pharmacognosy, the study of naturally occurring compounds – particularly those from plants – that can be used medicinally and in other ways, draws both upon biology and chemistry. When it became recognized as a separate discipline in the early part of the 18th century, it focused on the identification, preparation and commercialization of drugs, the majority of which came from plants at that time. From this practical foundation, pharmacognosists have developed sophisticated techniques which allow them to isolate active principles, determine their chemical structure, understand the way they are produced in living organisms and show the effect they have on people. These researchers also explore ways of producing greater quantities of natural drugs through tissue cell culture, genetic manipulation and other technical methods.

Ethnopharmacology places greatest emphasis on describing the medicinal properties in remedies used by local people. It also focuses on how they select, prepare and administer these curative plants and animals. This approach, which combines perspectives from chemistry, botany and anthropology, requires ethnopharmacologists to divide their time between the laboratory and the field. In order to select plants for chemical analysis, they interact with specialist plant users, such as herbalists, hunters and fishermen, who have developed an empirical knowledge of chemical compounds by tasting and smelling plants, observing the effects of medicinal herbs on patients and noting the link between the morphology and the curative powers of plants. Because of their close link to local people, some ethnopharmacologists seek to apply their knowledge and discoveries to improve health conditions in the communities in which they work.

Although the study of medicinal properties attracts the most attention, some researchers focus on other chemical constituents found in biological organisms. **Nutritional chemists** measure the quantities of specific nutrients found in edible animals and plants in order to discover the extent to which they fulfil dietary requirements. They may focus on the traditional foods consumed by local people, carrying out analyses of wild and managed edible plants not studied previously by other nutritionists. **Chemical ecologists** study how biological compounds affect the relationships between plants and animals, with special emphasis on plant–insect interactions. This research has inspired some ethnobotanists to explore how human diet and medicine have evolved in response to plant chemicals and how the concentration of these chemicals becomes modified in plant populations managed by people.

Whereas pharmacognosists and ethnopharmacologists focus on health and nutrition, **natural products chemists** work on a broad range of biological compounds that includes latexes and resins employed in industrial processes, essential oils used to make perfumes and other substances that have diverse applications. Chemists with a particular interest in agriculture focus on natural insecticides, as well as on the macronutrients found in plants used as a 'green fertilizer' to enrich the soil. Other researchers study hallucinogenic and narcotic substances employed in ritual or recreation as well as naturally occurring poisons which people use for hunting, fishing and other activities.

Natural products chemists, pharmacognosists and ethnopharmacologists are primarily interested in plants, but they also work on marine organisms, insects, animals and soil bacteria. Although all focus on naturally occurring chemical compounds, these researchers exhibit different degrees of breadth in their academic formation and practical experience. Pharmacognosists and ethnopharmacologists study biology and chemistry, whereas natural products chemists focus primarily on chemistry and tend to have less background in the social and natural sciences. Ethnopharmacologists usually gain extensive practical experience in anthropology and linguistics during their fieldwork, while pharmacognosists and natural products chemists typically spend their time in the laboratory and consequently have less understanding of the cultural context of the use of plants and animals by local people.

3.1.3 Sampling methods

Before setting out to collect medicinal plants for laboratory analysis, you should decide which sampling method you will follow. This decision is particularly important when working in the tropics, where both plants and secondary compounds are more diverse than in temperate zones.

Some research projects call for **random sampling**, taking any plant which you can collect in sufficient quantity and quality. The benefit of this approach is that you will be able to sample a diverse range of plants and make a large number of collections rather quickly. The disadvantage is that you will probably take in a relatively low proportion of species that show biological or pharmacological activity. Random sampling is used only by researchers who have the budget and facilities to screen large numbers of species. The results of these projects can be of great interest to ethnobotanists, because they reveal not only which locally-used medicinal plants contain chemical compounds that are physiologically active but also the overall proportion of plants with potential biological activity that are actually being used by local people.

How can we increase the proportion of plants that screen positively, thus saving research time and money? One way is to follow a **chemotaxonomic approach**. Specific secondary compounds, such as certain flavonoids, are often restricted in distribution, being found only in groups of related plants. For example, types of flavonoids known as isoflavones are common in species of the *Fabaceae*, but are found in few other plant families. Of the over 5500 types of alkaloids known, many are confined to a single genus or subfamily. Only a single alkaloid has been found in the many species of *Bombacaceae* tested thus far, but the *Solanaceae*, *Rubiaceae* and *Ranunculaceae* are the source of hundreds of distinct forms.

You can increase the efficiency of your sampling by knowing the distribution of the plant chemicals you wish to study. If it is known that a certain terpenoid is present only in the *Asteraceae*, you can limit your collections to this family. If you find an interesting type of alkaloid in one species of the *Rubiaceae*, you can proceed to gather closely related species, then collect more distantly related species of the same genus and later broaden your sampling to other genera of the family.

A second way to increase your rate of success is to use the **ethnopharmacological method**, choosing plants which are used as medicine by local people. This is the approach taken by most ethnobotanists, because it is the most efficient way of testing if the safety and effectiveness of local medicines can be corroborated through phytochemical and pharmacological analysis.

There is often significant agreement between the ethnopharmacological and chemotaxonomic methods of selecting plants. John Brett [32], in his study of Maya medicinal plants, discovered that certain plant families known to have large numbers of pharmacologically active chemicals – such as the Solanaceae, Asteraceae, Rosaceae and Ranunculaceae – are over-represented in the Maya pharmacopeia as compared to their general abundance in nature. Similar correlations have been pointed out in other parts of the world, which means that the two methods could be mutually reinforcing in many areas.

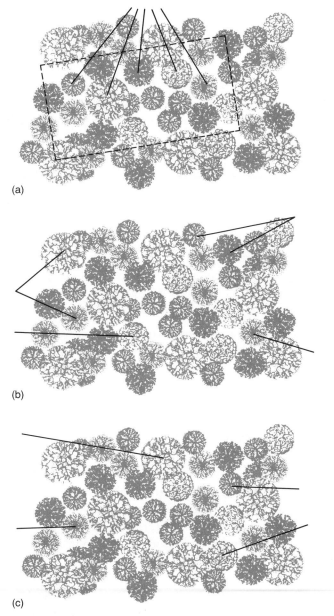

Figure 3.2 Representation of three strategies for selecting plants for screening, as visualized by Michael Balick, based on his research in Central America: (a) random collection; (b) targeted plant families or chemotaxonomic collection; and (c) ethnopharmacological collection.

Figure 3.2 illustrates the three approaches and Box 3.2 highlights an empirical comparison of the random and ethnopharmacological methods of sampling.

Box 3.2 Comparing sampling methods

Michael Balick, director of the Institute of Economic Botany (IEB) at the New York Botanical Garden, has begun an empirical assessment of the efficacy of different methods of sampling plants for phytochemical analysis [33]. He and his colleagues are working with the National Cancer Institute (NCI) of the United States to attempt to define plants which show potential in treating AIDS and certain types of cancer.

The first step in this project was to make bulk collections of 1500 species in Latin America and the Caribbean for laboratory analysis. Because there are an estimated 110 000 plant species in the New World tropics, it was necessary to use a sampling strategy that guaranteed the highest possible number of 'hits' – positive reactions against the AIDS virus. Based on collections made in Honduras and Belize, Balick set up a contingency table comparing the activity of those plants collected randomly with those selected using the ethnopharmacological method (Table 3.1). He found that five of 15 plants chosen with the ethnopharmacological method showed initial activity versus one of 17 species selected randomly.

Table 3.1 A comparison of activity in the *in vitro* anti-HIV screen of plants collected using the random and ethnopharmacological sampling method. $p = 0.10$

	Random	Ethnopharmacological
Active	1	5
Inactive	17	15

He analyzed the results with the Fischer exact probability test, which shows if there is a statistically significant difference between two or more sets of observations. This test yielded a probability of 0.10, a level that most statisticians would consider barely sufficient to reject the null hypothesis that there is no difference in activity between the two sets of plants. Although inconclusive, these preliminary results provide incentive to continue selecting plants that are used locally and to compare their activity with plants that are sampled by other methods. In a future stage of the project, Michael Balick and his colleagues plan to take a chemotaxonomic approach to selecting plants and to compare the efficacy of this method with random and ethnopharmacological approaches.

The ethnopharmacological method is particularly useful when searching for compounds which are effective in treating specific health conditions. For example, when Brazilian ethnopharmacologists Elaine Elisabetsky and Zuleica Castilhos set out to discover new botanical sources of analgesics, or pain relievers, they first designed a questionnaire to elicit various Brazilian plant remedies used for stomachache, headache, rheumatism and other types of pain [34]. When Michael Heinrich, a German biologist and anthropologist, was looking for Mexican plants used in the treatment of malaria and gastrointestinal disorders, he started by asking local people which plants they use against recurrent fevers and for stomach ailments such as diarrhoea.

This approach may be strengthened by comparing the plants used by different ethnic groups who live in the same region. It may be that species used in a similar way by a majority of local peoples are more likely to contain a physiologically active substance than species which are used by only one ethnic group or in a single community. Many researchers argue that commonality of use, whether arising through independent discovery or interaction between people of different cultures, is directly related to the degree of effectiveness of a remedy [35, 36].

This hypothesis should be tested on a broad range of plants in multi-ethnic regions before it is accepted as an absolute rule. In any culture, there are plants which are selected because they are strongly symbolic of the disease they treat, not because of the active principles they contain. For example, twining plants are often employed to treat snake bite, a use governed by the morphology of the plants and not their chemical composition. Frequency of use is also affected by the availability of the plant remedy and by the prevalence of the disease which it treats [37].

A corollary of the ethnopharmacological method is employed when collecting other types of useful plants. For example, if you are interested in analyzing the link between the nutritional composition of plants and human diet, choose species that are consumed by local people. Similarly, if you wish to measure the levels of macronutrients and micronutrients that certain species contribute to agroecosystems, collect plants which are being used as mulch and fertilizer by local farmers. Two examples of these studies, carried out by chemical ecologist Timothy Johns and his colleagues, are given in Box 3.3.

Box 3.3 Assessing non-medical uses of plants: the value of edible greens and 'green manure'

Timothy Johns, a professor of human nutrition at McGill University in Quebec, Canada, has carried out numerous ethnobotanical and phytochemical studies in Latin America and Africa. This research has focused on the relationship between the biological and the cultural aspects of human diet, health and agriculture [38, 39].

In a study carried out with Sarah Booth and Ricardo Bressani, Timothy Johns assessed the nutrient content of leafy vegetables consumed by Kekchi-speaking people in Alta Verapaz, Guatemala. For each analysis,

---

5–20 plants (approximately 200 g fresh weight) were collected from household gardens or maize fields. These edible greens were gathered with people from the community, who indicated which plant parts to select and in which localities the plants could be found. Additional plants were purchased in local markets, when possible. When sufficient material was available, an additional 200 g sample was collected so that a woman from the community could prepare it in a typical way, boiling the greens in an aluminum pot over an open fire, using local water. Once the fresh and cooked collections were weighed and sealed in plastic bags, they were placed in a portable cooler and transported to Guatemala City, some 3–4 hours from the research site. The samples were stored at −20°C at the Institute of Nutrition for Central America and Panama (INCAP), where standard laboratory procedures were employed to reveal the composition of each plant, including moisture, total fat, crude fiber, protein, carbohydrates, minerals (calcium, phosphorus, iron, potassium and magnesium) and carotene (a source of vitamin A).

Of the 13 species analyzed, the nutrient composition of five had never been previously reported. The study confirmed the importance of non-cultivated edible greens as sources of protein, minerals and vitamin A, substances which are often deficient in the diets of impoverished rural people. Results such as these are important to INCAP and other organizations which seek to alleviate malnutrition and diet-related illnesses in Central America.

In a separate project, Timothy Johns joined forces with Alejandro Camino, a Peruvian anthropologist, to study a fern used as a fertilizer in Andean agriculture. In many regions of the world, efforts are made to increase soil fertility by adding plant mulch to cultivated fields, a practice which is called **green manuring**. In the Andes, *laki-laki* (the Quechua name for *Dennstaedtia glauca*), is gathered in the wild and placed on fallowed fields just before they are cultivated. The fern fronds, used in combination with sheep manure, are left to dry in the field and are eventually dug into the soil. Chemical analysis of the fern showed high levels of nitrogen, phosphorus and potassium, making it superior to animal manure as fertilizer. Working as a natural recycler, *laki-laki* takes these macronutrients from soil. It anchors itself with large rhizomes along rivers and in humid ravines, where it efficiently absorbs the nutrients lost from steep cultivated fields through leaching and runoff. When the fronds are placed on fields, these nutrients are restored to the soil. The phytochemical study by Camino and Johns allowed the benefits of *laki-laki* to be quantified, corroborating the utility of this aspect of traditional agriculture.

3.2 Screening

3.2.1 Advantages to field screening

There are two ways of going about screening. You can go to the field with the necessary equipment to test the samples yourself or you can send them to a pharmacognosist who has agreed in advance to process the material in his or her laboratory.

Field testing enables ethnobotanists to identify plants which merit extensive investigation in the laboratory, allowing bulk collections of promising species to be made immediately. Another advantage of field screening is that the plant can be tested fresh, rather than in a dried or extracted form. This is important if you are working with plant compounds that are apt to disappear from the material before it reaches the laboratory.

Drawing upon techniques that are used in laboratories worldwide, pharmacognosists have developed portable kits which enable screening for plant chemicals to be carried out in the field, even when electricity and running water are lacking. The procedures are relatively simple to learn and no specialized training in chemistry or botany is necessary. It is sufficient to test a few leaves or other plant parts, usually weighing no more than 10–50 g. The results are visual, because the presence of a given compound is indicated by a solution turning color or forming a precipitate, a solid material that forms within the extract.

If you wish to screen plants in the field, consult first with an ethnopharmacologist about the basic glassware and reagents you will need and the exact procedures you will perform. He or she can show you in a matter of hours how to carry out some simple tests that are the first step towards detecting alkaloids, saponins and other compounds.

It is possible to screen many dozens of plants in a few days, especially when there is a team of researchers working in coordination. For example, the Malaysian Natural Product Research Group holds occasional excursions to selected natural areas where it carries out extensive field testing of plants. In 1986, it selected the Danum Valley Conservation area as the site for its third phytochemical screening project [40]. This conservation area, which is situated in the central part of Sabah, includes 438 km^2 of some of the last remaining undisturbed mixed dipterocarp forests in Borneo. A group of chemists and botanists from Malaysia and abroad stayed in a research center for one week and worked intensively in a 1.5 km^2 sample area. They made herbarium vouchers, collected random samples for phytochemical analysis and screened for alkaloids. They later carried out tests for other compounds in the laboratory. Of a total of 223 species screened, 46 tested positive for alkaloids, 91 contained saponins and 59 had steroids and triterpenes. Some of these plants were then selected for more intensive study.

3.2.2 Drawbacks to field screening

Despite the simplicity of using these techniques, there are disadvantages to field screening for chemical compounds. Although relatively easy to carry out, field testing leaves less time for other aspects of an ethnobotanical inventory. The funds spared for buying reagents and the hours dedicated to preparing plant samples and recording results might better be spent in making voucher specimens or interviewing local people. Remember that you can always analyze samples later or send them to a qualified colleague for testing but once you leave the field you will have no opportunity to collect plants or speak with local people.

Another disadvantage of field testing is that the field techniques are not very precise. You only discover the presence of groups of natural products such as terpenoids or alkaloids, without identifying the precise compounds. Some plant compounds, present in extremely small quantities, are missed by the field test but would be detected in a more sophisticated laboratory screen.

Screening is of little scientific merit in itself. The presence or absence of a chemical compound does not allow us to conclude that a plant is effective for the purpose given by local informants. This can only be established after extensive bioassays, tests on animals and clinical trials which must be carried out in a laboratory under strict controls. Keep in mind that most phytochemists, who view screening as simple preparatory work for more sophisticated analysis, do not bother to publish the results of their phytochemical surveys. Although this information may be presented in student theses or in regional conferences, it is not considered of sufficient rigor or interest to be submitted to scientific journals.

3.2.3 Pharmacological and biological screens

Apart from phytochemical screening, there are many types of pharmacological screens, most of which have to be carried out in a well-equipped laboratory. For example, researchers at the Central Drug Research Institute in Lucknow, India use 82 separate screens, including ones for activity against specific bacteria, fungi, protozoa, intestinal worms, viruses and spirochaetes which cause disease [41, 42]. They also probe for effectiveness against specific health problems, such as cancer and inflammation, and they judge the effect of the compound on different physiological and anatomical systems such as reproduction, digestion and circulation, among many others.

Most of these pharmacological and biological screens require access to a relatively sophisticated laboratory, but there are two simple techniques – the **brine shrimp screen** and the **antibacterial screen** – which can be carried out with rudimentary equipment in simple laboratories. Although some researchers feel these tests have limited usefulness and little scientific relevance, they demonstrate some of the basic principles involved in pharmacological screens.

Brine shrimp (*Artemia salina*) are small aquatic animals that can be grown in a special solution which resembles saltwater from the sea. In order to test the cytotoxicity potential of a plant – and thus its probability of containing an anti-cancer agent – measured amounts of plant extract are added to various containers holding known numbers of shrimp in solution. After 24 hours, the surviving shrimp are counted, allowing the calculation of a cytotoxicity measure called the **LD-50 value**. This corresponds to the concentration of a compound in solution which kills 50% of the brine shrimps.

The biological activity of plants against numerous types of bacteria can be tested in a different kind of screen. The bacteria are grown on a medium called **agar**, a gel derived from various algae, which is poured into round glass dishes called **petri dishes**. Measured amounts of a plant extract are placed on paper discs, which are then set on the surface of the bacteria-inoculated agar. All of these procedures must be done under sterile conditions to ensure that no undesired bacteria are introduced onto the plates. After a period of 18–24 hours during which the plates are **incubated** – left in a warm place that stimulates the bacteria to grow – **clear zones** or bacteria-free circles can usually be observed around some of the paper disks, indicating that the plant extract has inhibited the microbes. The size of the circle is related to the concentration of the extract, which means that, in order to compare the activity of different plants, a similar amount of extract from each species must be used. Box 3.4 explains how the brine shrimp bioassay and antibacterial screening were used to test plants gathered in a national park in Puerto Rico.

Box 3.4 The biological activity of plants in the Caribbean National Forest

The Caribbean National Forest comprises some 28 000 acres of tropical rain forest in the northwest part of Puerto Rico. There are an estimated 3000 species of higher plants in the park, including a number of endangered and endemic species.

Between 1987 and 1989, Ricardo Guerrero and Iraida Robledo of the University of Puerto Rico examined the biological activity of 113 botanical species that were randomly collected within the park boundaries [5]. The collections represented 54 botanical families and included 19 species which were endemic to the island. A voucher specimen of each collection was placed in the herbarium of the University's School of Pharmacy.

In order to make an extract of each plant, 20–30 g of dry material was ground and then mixed with 95% ethanol in a blender. After 24 hours, the suspension was filtered and evaporated under vacuum at 40°C. 50 mg of each extract were then dissolved in 5 ml of methanol.

Under sterile conditions, 100 µg and 500 µg of each extract were applied to discs of filter paper which were then placed in petri dishes inoculated with a single type of bacterium. In all, the extracts were tested against six different types of bacteria related to those that cause disease in people. After these plates were incubated for 18 hours at 35 °C they were examined to see which of the plant extracts had inhibited which strains of bacteria. Inhibition was judged by measuring the size of the clear zone, a bacteria-free circle that forms around the discs. If the paper disks impregnated with either concentration of extract showed clear zones of 7 mm or more, the researchers considered that the bacteria were susceptible to inhibition by the extract, whereas clear zones of 6 mm or less were taken to indicate that the bacteria were resistant to the extract. The screen revealed numerous plants that were active against the class of bacteria designated as Gram positive. Gram negative bacteria were in general resistant to the extracts. Two plants showed a particularly high level of activity. At a high concentration, *Turpinia occidentalis* (*Staphylaceae*) inhibited five of the six bacteria tested and *Clidemia hirta* (*Melastomataceae*) showed activity against four of them.

In a separate test, brine shrimp eggs were incubated in artificial sea water called Instant Ocean. After hatching, 10 brine shrimp were placed in each of a series of small plastic containers. Different concentrations of plant extract were then added to one container of shrimp, while another identical one (the control) received only methanol, in the amount included in the other samples. After 24 hours, the number of live shrimp was recorded. These experiments yielded data that permitted the researchers to calculate the LD-50 of each extract, which is the dose that kills 50% of the organisms. Three species of *Euphorbiaceae* were found to be toxic to brine shrimp at very low concentrations.

After achieving these results, Ricardo Guerrero and Iraida Robledo searched the phytochemical literature for more information about the active plants. They found some studies on the same genera, but none on the particular species that had shown strong activity in the screens. They recommended that more sophisticated bioassays be conducted on the promising species and that the active principles be isolated and further tested.

The same criticisms made of field testing for chemical compounds can be leveled at employing brine shrimp, antibacterial and other simple biological screens. These tests yield relatively little information compared to the sophisticated assays that are done in fully equipped laboratories. Furthermore, these screens can only detect a limited range of biological activities. Peter Waterman, a phytochemist at Strathclyde University in Scotland, has pointed out that some important plant chemicals in use today, including morphine, are not toxic to brine shrimp and would not have been detected in this bioassay. This is because the brine shrimp test is specifically used to measure cytotoxicity and does not indicate the general activity or toxicity of a plant extract.

There appears to be a general consensus among ethnopharmacologists that field screening is best reserved for educational purposes – such as giving local people and colleagues an introduction to the procedures followed in laboratories – or for initial testing of plants as part of a broader phytochemical research project. It is best not to carry out screening as part of your research project unless you have already forged links with pharmacognosists who have the time and ability to carry out sophisticated analyses of selected plants, including isolation of pure compounds, structure elucidation and testing for biological activity.

This is the approach being contemplated in Kinabalu Ethnobotany Project. Local collectors are preparing 50–100 g samples of the useful plants they are collecting in their communities. These will be screened in park headquarters by project coordinators in collaboration with students from various Malaysian universities who come to gain practical experience in phytochemistry and ethnobotany as part of their thesis work. The students will work in a simple laboratory housed in the Natural History Museum of Kinabalu Park, where they will prepare extracts of the plant samples and carry out initial isolation of plant compounds. A network of Malaysian and Southeast Asian pharmacognosists and ethnopharmacologists who are collaborating on the project will carry out further work on promising plants in their laboratories, including testing for biological activity and structural elucidation of selected compounds.

If you embark on a similar initiative, remember that pharmacognosists are as overwhelmed with phytochemical samples as taxonomists are with voucher specimens. It is easier to collect or screen a plant than it is to identify it or analyze its chemical composition. Of the many hundreds of medicinal plants that you detect in a ethnobotanical inventory and screen in a phytochemical survey, few if any will be studied extensively in the laboratory.

3.3 Collecting plants for phytochemical analysis

After analyzing the results of a screen, you may wish to make bulk collections of promising species which merit being analyzed with technically sophisticated methods in a fully equipped laboratory. There are many similarities between collecting plants for an herbarium collection and for phytochemical analysis. The major differences are that, in addition to making a voucher specimen, you must collect a larger amount of material from each species designated for the laboratory and that this sample must be prepared in a way that alters as little as possible the chemical composition of the plant.

In the following discussion, I follow the order used in Chapter 2 in the discussion on how to make herbarium specimens, but focus on the special techniques used to collect ethnopharmacological samples. It will be useful to refer back to the section on general plant collecting for additional details on many of these steps (p. 28). The following discussion focuses on the collection of medicinal plants, but applies as well to the sampling of other useful plants. Keep in mind that these are general guidelines which may be modified by the laboratory researchers who are collaborating on your project.

3.3.1 Selecting the locality, time of day and population of plants

When selecting the species that you wish to test further, consider whether or not the plant populations will yield a sufficient quantity of material for analysis. Avoid collecting rare species or substantially diminishing a stand of plants that is being used by local people. Several techniques for harvesting bulk quantities of plants in a sustainable way are described in Box 3.5.

Box 3.5 Sustainable harvesting of bulk plant samples

As mentioned in Chapter 2, you should be concerned about over-collecting rare plants when you make voucher specimens for botanical identification. If you plan to gather bulk quantities of plants for phytochemical analysis, you must pay even greater attention to your methods of sampling and collecting to avoid over-harvesting certain species [43]. These measures are important to protect supplies for local resource users as well as for ensuring conservation of genetic diversity.

Whenever possible, do some background bibliographic research on the species, focusing on its distribution, abundance and reproductive biology. Ensure that the species you have selected are relatively abundant in the collecting area. Choose large and mature plants that can better withstand the impact of harvesting than small or immature ones.

Take all necessary measures to guarantee that species can continue to produce new vegetative growth or seeds after being harvested. If you must gather the whole plant, ensure that some individuals remain in each stand. When picking leaves from large herbaceous plants and shrubs, leave enough foliage so that the plants can continue to grow and reproduce. On trees or shrubs, harvest from side branches rather than the main trunk and branches. If you wish to collect flowers or fruit, leave enough reproductive material to ensure that there is a sufficient seed set for production of new seedlings in coming years.

Bark and root harvesting require special care. As indicated in Figure 3.3, bark should be stripped from opposite sides of the trunk or branch, allowing the tree to survive and the bark to regrow eventually (but note that species of tree vary in their ability to regrow bark). Many root-like organs, such as rhizomes, stolons, tubers and bulbs, are actually modified stems. These often have buds which allow vegetative reproduction even if some part of the root is harvested. Woody plants that have complex root systems can survive if most of the roots are left intact. Harvesters should leave a large portion of the bark and the root system to stimulate regeneration of the species. Cutting

---

down trees, girdling trunks or removing a large amount of the roots could threaten the plant population, potentially depriving local people of a resource which they value.

Figure 3.3 Methods of collecting bark samples. In (a) the trunk or limb is girdled, causing the tree or branch to die, whereas in (b) the bark is stripped from opposite sides, allowing the tree to survive.

If you expect to need a large supply of a particular species over a number of years, experiment with various methods of sustainable harvest to see which is the most effective. Some plants can be brought under cultivation and the abundance of others increased through various horticultural or sylvicultural techniques.

Be aware that individuals of the same species growing at different localities may have strikingly different levels of active compounds. Different parts of a plant will vary in their chemical composition and samples collected at various times of the day and night and in different seasons may also fluctuate in potency. If you are employing the ethnopharmacological method, follow the local people to the plant population they harvest and collect at the same time as they do. When following a random or chemotaxonomic method, attempt to make separate samples of plants growing under diverse environmental conditions and at different stages of growth.

Ensure that you are collecting only one type of plant and not mixing material from various species. Even when an herbal remedy is composed of many different species of plants, it is important to make a different collection of each one. Collect and prepare each type of plant part separately, even when they come from a single species. The chemical content of leaves, roots, fruits and bark may not be the same and these organs will take different amounts of time to dry.

The plants should not be infested with bacteria, fungi or small insects, because these organisms will alter the results of the phytochemical analysis. Levels of certain compounds, such as alkaloids and terpenoids, will rise in response to attack by insects or herbivores. If the local people are specifically collecting infested material for medicinal or other purposes, you should do the same. Otherwise, strive to make collections from plants that are healthy and whole. Whenever possible, choose a population that contains some flowers or fruits, allowing you to make a fertile voucher specimen that will facilitate identification and serve as a permanent record of the material tested.

3.3.2 Collecting the plants

Phytochemists need varying amounts of dry material to carry out their studies, sometimes up to a kilogram or more. Fresh plants contain a large amount of water, sometimes amounting to 80–90% of the total weight of a sample of leaves or fruits and 40–50% or more for most barks and roots. Plan on collecting at least 3–10 times as much fresh plant material as you will need of dry material, depending on the succulence of the plant.

In the laboratory, the dry plant material will be ground into a powder with the help of a mortar and pestle or a mill. In the field, you can harvest the plants in any way that facilitates the preparation and shipment of the material as long as it does not decimate the local population of plants or reduce its utility to local people. As you are harvesting the plants, remember to leave enough fertile material to make the number of voucher specimens that you require.

3.3.3 Preparing the plants

Once you have assembled a sufficient quantity of material, you should break or cut the plant parts into a size that will make them easier to preserve and handle. Roots and bark can be chopped into pieces using clippers or whatever local instrument is used for clearing vegetation. Fruits, depending on their consistency, are cut into pieces or simply left whole. Leaves are left intact or cut into pieces with a knife, according to their size and texture. The exact way of preparing the plants is based on how you plan to preserve them.

3.3.4 Preserving the plants

There are several ways to deliver the plant material to the laboratory: (1) fresh, (2) frozen, (3) dried or (4) preserved or extracted in alcohol [44, 45]. When arranging collaboration with a phytochemical laboratory, enquire how the chemists would like to receive the material.

Unadulterated, fresh material is desirable because all methods of preservation cause some active principles to break down, evaporate or otherwise disappear from the plant. Yet it is usually impractical to handle fresh plants, because plants often wilt, become desiccated or become moldy before they can be delivered to distant laboratories.

An alternative (even more impractical and expensive than sending live material) is to freeze the plant parts and deliver them packed in dry ice or liquid nitrogen, materials which are often hard to obtain. In the few cases where unadulterated material is required, the fresh or frozen plant parts can be packed into an ice box or similar receptacle which is cool, dark and moist.

Most pharmacognosists accept material preserved in a conventional way, typically dried. Keep in mind that drying affects the chemical composition of plants and that recommendations can vary according to the specific compounds of interest and the procedures of extraction and isolation that will be followed.

Dry material is easy to handle and relatively inexpensive to ship. Rapid desiccation traps most non-volatile compounds and even many substances which evaporate readily. Slow drying increases the chances that compounds will be broken down by enzymes or that water remaining in plants will cause the material to become moldy.

Ensure that the plants are cut into relatively small pieces that will dry out quickly. The moisture can be evaporated quickly by placing the plant fragments in the sun or over a screen in a plant dryer, but this also heightens the risk of losing substances which evaporate easily or which are damaged by exposure to light. If you choose to place the plants in a dryer, keep the heat low and monitor the temperature with a thermometer, ensuring that it stays under 50°C. Some compounds require even lower temperatures.

Drying in the sun is an acceptable method, but keep in mind that it is difficult to calibrate the intensity of the heat and that the radiation could affect the chemical composition of the plants. Drying in the shade is the preferred method of preparing plants from which volatile essential oils will be extracted, but you should monitor the material carefully to ensure that they do not turn moldy, especially in humid tropical zones. In all the above techniques, the plants are turned several times a day to ensure that drying is uniform.

Preserving plants in alcohol can be done as described in the section on plant collecting. Place the plant pieces in a durable plastic bag or container and cover them with a sufficient quantity of alcohol. Keep in mind that bulk collections saturated in alcohol will be heavy and difficult to handle.

If you wish to extract your plants in the field, the most common method is to chop the material into pieces, place them in a container and add methanol or ethanol. Extraction in the field is practical only if you are planning on testing a few plants. Most ethnobotanists prefer to concentrate on collecting bulk quantities of the plant in the field, leaving the extraction to colleagues who work in a laboratory.

3.3.5 Keeping a field notebook

Although the field notebook for phytochemical collections includes the information that is recorded for any ethnobotanical voucher, there are a few special points to keep in mind. Use the same collection number for both the herbarium voucher and the bags containing material for phytochemical study. This ensures that the two are inextricably linked and that no confusion can result about the origin or identity of the material processed in the laboratory. Remember to note down the specific information required by the phytochemists who will analyze the plants: (1) how the plants are preserved, including the solvent used or temperature and duration of drying; (2) any special environmental conditions of the collection locality; (3) the maturity of the plant, including its stage of flowering and fruiting; and (4) the time of day that the collection was made.

Robert Toia, a chemist at the University of New South Wales in Australia, suggests paying attention to several attributes of the fresh plants that will not be evident in dried specimens [46]. Look for latex or colored sap. Crush a few leaves and note if the smell is particularly aromatic, indicating the presence of an essential oil or other fragrant compound. Scrape the bark off the trunk, stem or root of the plant and look for a yellow to red coloration which may result from the presence of quinones and similar substances. Following the advice of local people, you could even place a leaf or two in your mouth, tasting for the sweetness, bitterness or sourness given by distinct classes of plant compounds. Exercise caution in tasting plants, because some species are highly toxic. As a simple rule, avoid putting any plants with latex in your mouth. Record all of your observations in the field notebook.

In the course of your fieldwork, make detailed notes on local methods of harvesting, preparation and use of plants. Elaine Elisabetsky has stressed that understanding the cultural context in which plants are used can guide phytochemical analyses [47, 48]. She encourages full documentation of several aspects of medicinal plant use. The **preparation** should be recorded, including whether or not the plants are heated or if they are mixed with water, alcohol, fats or other substances which act as solvents, making active compounds more easily absorbed by the body. The **mixture** of different plants in a single remedy should be noted.

Elisabetsky states that when distinct species are combined, there may be chemical interactions between them that produce the healing effect. Alternatively, a single species may be responsible for the therapeutic action, while the others are secondary in importance. The **posology** – defined as the dosage and way of administering the remedy – should be described. The quantity of plant material given is often carefully calculated by traditional curers, as is the length of time during which the herbs are taken. A critical element of the posology is whether the plant is administered internally (taken into the body) or externally (applied outside the body in the form of a massage or a bath). Finally, you should have a good grasp of local concepts of illnesses, their symptoms, their indigenous names and how they correspond to Western disease categories. Although many health conditions correspond to diseases recognized by Western medicine, be aware that some, called **folk illnesses,** may be difficult to describe by comparison to diseases and symptoms recognized by Western doctors.

Edward Croom, Jr, who works in the Research Institute of Pharmaceutical Sciences of the University of Mississippi, has included these recommendations and others in a set of guidelines devised to aid researchers who record the use of medicinal herbs and send material for phytochemical analysis [49]. Some of his suggestions for documenting and evaluating herbal remedies are reproduced in Table 3.2.

3.3.6 Labeling the specimens

Standard collection labels should be made for the herbarium voucher specimens, as described in the section on botanical collecting. Each sample for phytochemical analysis should be clearly marked with the collection number and should be accompanied by a complete account of the locality data and ethnobotanical notes about the specimen. This information will be essential to guide the work of the phytochemist, but does not need to be produced as printed labels. A rough print-out from a computerized database or a photocopy of the relevant pages of the ethnobotanical notebook is sufficient.

3.3.7 Shipping the specimens for analysis

If the material is completely dry, it can be packed into plastic bags. Keep in mind that even traces of moisture can cause contamination or molding of the plants if they are completely sealed in plastic. For samples that retain some ambient moisture, use bags made of burlap or another material that allows air to circulate.

The shipping notice should indicate that the specimens are for phytochemical analysis and should specify the mode of preparation. If the samples are to cross an international border, ensure that the proper phytosanitary permits have been obtained and are clearly displayed on the box.

The herbarium voucher specimens are usually shipped separately from the

Table 3.2 A checklist of information suggested by Edward Croom Jr for recording information when collecting medicinal plants for phytochemical analysis

---

Information on the medicinal plant

Collection
    Part of plant collected
    Preferred collection time, stage of development and location
    Special procedures
    Storage of plant

Extraction and processing
    Plant(s) used
    Type and amount of plant part(s) used
    Type and amount of solvents used
    Type and amount of heat applied
    Length of processing
    Equipment
    Form of remedy (powder, decoction, etc.)
    Storage

Drug therapy
    Amount and timing of dosage
    Route of administration and method of use
    Disorder(s) treated
    Therapeutic activity, including informant's assessment, personal observations, medical records and information from the scientific literature
    Status of use (e.g. former or current)
    Adjunct therapy (e.g. diet, sweat baths, prayers)

Information on community health characteristics

Medical beliefs and setting
    Practitioners, range of specialties, selection, training, competence
    Diagnostic tools, techniques and criteria for illness
    Criteria for cure of an illness
    Cause(s) of an illness
    Therapeutic setting, including location, objects and people present

Community characteristics
    Major causes of death and disease
    Nutrition level
    Sanitization level
    Standard of living

---

phytochemical samples, because in most cases the two sets of materials will be handled by different researchers or institutions.

When the material is received at the laboratory, it will be further processed. Phytochemists try to analyze the material as soon as possible but are often overloaded with samples. If you wish to send fresh or frozen material, be sure that your colleagues are able to receive and process it immediately. Until it can be extracted, preserved material is stored in a room which is controlled for moisture, light, temperature and insects.

Before extraction, some phytochemists make a further voucher specimen, keeping one or two hundred grams in a glass jar or metal box. If there is ever any doubt about what has been analyzed, this material can be cross-checked with the herbarium voucher specimen to verify its identity. In order to ensure that this step is followed, you can send a small bag of the sample marked 'phytochemical voucher: do not process' along with the bulk collection.

3.4 The ethics of searching for new plant products

Before you begin an ethnopharmacological project, consider the ethical implications. At present, there is much debate about how the benefits from the marketing of new products derived from living organisms should be shared [50–52]. Such products may be manufactured directly from materials in organisms, or be modified from compounds obtained from organisms, or be synthetic, but with their discovery based partly on studies of natural compounds in organisms. Many of the enterprises capable of developing plant-derived pharmaceutical and industrial products are in the developed world, yet the most promising botanical species are found in tropical, developing countries. Within these countries, it is often ethnic minorities or rural people whose knowledge stimulates the selection and screening of specific plants.

Commercial enterprises must recognize the need to share benefits, including profits, with countries that are the source of plant genetic resources. In addition, local people must be compensated for their intellectual contribution to discovering novel, plant-derived commodities. Researchers, and everyone else involved in prospecting useful plants, share in the responsibility to ensure that the benefits get to the right people.

The general guidelines for conducting ethnobotanical research discussed in the final chapter provide some initial ideas about how to carry out ethnopharmacological work in an ethical way. A detailed reflection on this subject, written by Tony Cunningham, has been published by the World Wide Fund for Nature [53]. Early in the booklet, the author describes the problems that often surround collection of samples for phytochemical analysis:

> In the past, collecting has been largely uncontrolled, with sample material being taken for analysis to Europe, Japan or North America. Drug development and patenting has taken place without the knowledge of people in the country of origin, with no recompense for use of regional natural substances and often without any contractual obligation. When permit applications have been made, this has been done without indicating the commercial intent of the collectors. Local professionals such as botanists or foresters are

paid privately to collect samples for industrial companies or other sponsoring organizations. These people often do not understand the full implications of their work and their payment bears little relation to the potential value of the resource. As many developing countries pay them relatively little and hard currency is hard to get, it is understandable that this occurs. Nevertheless, it has important implications for regional development and conservation. From the point of view of national and local interests, why should these sponsoring organizations be allowed to treat a country's natural resources as global common property, especially considering that the natural resources may later be privatized by the same organization?

In order to remedy this situation, Tony Cunningham has proposed a code of practice for the collectors, representatives of pharmaceutical companies and other enterprises, university personnel, governmental workers and others who are involved in prospecting for new products from plants and animals. As a beginning, he suggests the formulation of regional and national legislation to control the collection and export of biological materials. These laws should result from discussion among the various actors involved in trade and study of living organisms and should take into account provisions of international agreements on intellectual property rights and the use of biological resources.

Extraction, screening and more sophisticated techniques should be carried out in local laboratories and with local expertise whenever possible. This will stimulate capacity building at a regional and national level and will ultimately strengthen local control of natural resources. As part of this effort to build the infrastructure of in-country institutions, any agreements for supplying biological material or extracts should be made with reputable organizations rather than individuals.

Of greatest relevance to ethnobotanists who are embarking on a field project is Tony Cunningham's proposal for a code of professional ethics. This code would require that all local participants and researchers be advised of the scientific objectives of the project as well as any potential commercial applications. Local people working in the project would be compensated in an equitable way and relevant national or regional organizations would receive fair royalty payments. Researchers would respect all national requirements for plant collecting, such as acquiring permits and working with local counterparts. The code furthermore proposes that confidential information not be disclosed and that the anonymity of local participants be respected, if they so wish.

A set of specific contract guidelines proposed by the Asian Symposium for Medicinal Plant Species and Other Natural Products (ASOMPS) provides good advice for anyone involved in ethnopharmacological fieldwork. Published in the *Manila Declaration*, a statement produced after the 1992 ASOMPS meeting in the Philippines (54), they include the following minimum standards for agreements between countries which are the source of plants (usually developing countries) and those where technically sophisticated analyses are carried out (usually developed countries). The suggestions can be modified to fit the particular country and laboratories with which you are working.

- The amount of material collected for initial screening should not normally exceed 100–500 g (dry weight) unless specific permission is obtained.
- Payment should include all handling expenses.
- Where screening is carried out with the aid of a partner organization in the developed world, a minimum of 60% of any income arising from the supply of extracts to commercial organizations should be returned to the appropriate country organization.
- The country organization should receive a minimum of 51% of any royalties arising from external collaboration that result in marketable products. As fair royalties are normally 3–5% the national organization would expect to receive a minimum royalty of 1.5–2.5%.
- The country organization should not sign any agreements that give indefinite exclusive rights to any external party. Exclusivity should be limited to no more than a two-year period.
- Complete evaluation results should be reported to the supplying country within 6–9 months.
- If there is a threat of destructive harvesting, costs of sustainable harvesting or development of alternative supplies must be borne by the external organization.
- The contribution of research participants should be recognized through co-authorship of publications (unless anonymity has been requested).
- Where possible, screening of extracts should be carried out in the country of origin and assistance should be provided to develop this expertise wherever possible.

3.5 Bringing phytochemistry back home

There are many ways that you can contribute to building local capacity as part of a phytochemical approach to the study of useful plants. If there is interest in the community where you are working, you can work with local people to screen, extract and analyze plants and to understand the results of phytochemical studies, much like the experience in Kinabalu Park. Always seek to support the work of in-country pharmacognosists and students by providing them with samples, opportunities to work with you in the field and financial support, when possible.

In addition, you may make the results of complex analyses accessible to local people by providing common-sense recommendations about the safety and efficacy of medicinal and edible plants. These actions are exemplified in the experience of TRAMIL, a Caribbean work group described in Box 3.6.

Box 3.6 The work of TRAMIL in the Caribbean Basin

TRAMIL is an applied research program that focuses on popular medicine in the Caribbean region. It was initiated by researchers from various disciplines – anthropology, botany, pharmacology and medicine – who began working with communities in Haiti and the Dominican Republic. The project has been expanded to other parts of the Caribbean [55]. The roster of participants is constantly growing and includes not only researchers but also herbalists and other local people who are interested in primary health care.

Through a series of surveys conducted in urban and rural areas, the TRAMIL participants gather information on the plants which are commonly used as household remedies. They begin by preparing a list of the ten illnesses that most affect people in the region. They then interview local people, inquiring if these conditions are treated by medical doctors, local curers or by family members who employ household remedies. Complex diseases treated by doctors or curers are excluded from the remainder of the survey, which focuses on the medicinal herbs used by the general population for minor health conditions.

The informants give a list of medicinal plants that can be used to treat these minor illnesses and they describe how each remedy is prepared and administered. The TRAMIL researchers subsequently make voucher specimens, enabling them to identify each plant mentioned in the interviews.

The most culturally significant plants are selected by gauging the level of consensus between respondents. If more than 20% of the people interviewed in a single region give the same use for a particular plant, it joins the list of species to be further evaluated by TRAMIL. If no species achieves this 20% agreement, the plants mentioned most frequently for the specific conditions are selected. Computerized databases and scientific articles are reviewed for information on the chemical composition, potential toxicity and therapeutic value of each of the selected species. In many cases, samples of the plant are analyzed in laboratories by pharmacognosists and ethnopharmacologists who participate in TRAMIL. Through these studies, researchers seek to reveal more about the constituents and biological activity of each remedy.

With this information in hand, TRAMIL collaborators seek to promote the use of plants which have been deemed effective for common ailments which can be treated at home, discourage the use of others which appear to be toxic and continue studying those for which information is still lacking. Every two years, a multidisciplinary seminar is held to appraise the progress of the work.

The fruits of each seminar are incorporated into an updated version of a book which contains information and advice on selected medicinal plants of the Caribbean. There is a line drawing of each species and written sections that give its geographical distribution as well as its common and scientific names, all of which help readers to recognize the plant. Uses drawn from the surveys and from bibliography on Caribbean popular medicine are reported and a review of the phytochemical literature and laboratory analyses accompanies the description of each plant. At the end of this detailed information, a summary recommendation is given based on a review of all the evidence by TRAMIL participants during week-long seminars. One of three ratings is given for each plant part used for each illness, as reflected in the initial surveys. (A) or 'TOX' means that the plant part is probably toxic when employed for the specific condition and its use should be discouraged; (B) or 'INV' suggests that the studies are inconclusive and the plant should be further investigated whereas (C) or 'REC' indicates that the plant part is probably beneficial and its use is recommended for the illness under consideration.

Through this book, TRAMIL seeks to communicate ethnopharmacological information on Caribbean medicinal plants to local doctors, pharmacists and other health professionals. A series of popular seminars and pamphlets provides similar information to the communities that have participated in the surveys and that depend on plants for their everyday health care.

4

Anthropology

Figure 4.1. Roberto Hernández López, a traditional medical practitioner from Santiago Comaltepec, harvesting the leafy shoots of *Cecropia* trees, which he combines with other herbs to treat diabetes. Field excursions with specialists are one approach to documenting local ecological knowledge.

4.1 Talking with local people

The initial ways to gather anthropological information are to talk with people, to watch what they do and to take part in their everyday activities. Although these ethnographic skills may seem so natural as to be taken for granted, they require proficiency and forethought if they are to yield reliable data.

Anthropologists have given special names to various field techniques [56]. **Participant observation** refers to living with people and sharing with them many facets of their life, from subsistence activities such as cooking, farming or gathering firewood, to ritual occasions such as marriages, religious celebrations or initiation rites.

Interviewing refers to asking people about their beliefs and lifestyles. In **open-ended or semi-structured interviews**, respondents give extensive responses to a series of general questions, some of which have been prepared in advance and some of which arise naturally during the course of the conversation. In-depth interviews may be held with a **local expert** or **key informant**, someone who has a profound knowledge of a particular aspect of local culture. Some anthropologists record the **life history** of these specialists, recounting the person's experiences as he or she grew up and attained expertise. These approaches, which are often referred to as **informal** or **qualitative methods**, yield responses that are used to write up general ethnographic accounts of the community and its culture. If the researcher wishes to analyze the responses with statistical methods, they must be coded and categorized before being interpreted.

Systematic or **structured interactions** involve asking a group of selected informants to respond to the same set of questions. These approaches, which are often referred to as **formal** or **quantitative** methods, yield verbatim answers that can be analyzed using various statistical methods without coding or categorizing the responses [57, 58].

Semi-structured and systematic interviews can give us good ideas of the ways people describe their lives and their natural surroundings, while participant observation allows us to see how people put their knowledge into practice. Let's say you are interested in how people use edible plants from their home gardens. In an open-ended interview, you might ask them how they select which species to cultivate, how the plants are harvested and how they are prepared for eating. A systematic interview could consist of asking informants to rank their preference for a number of edible plants that are grown in home gardens. In addition to these interviews, you might also observe directly the plants grown in gardens and the ways in which they are cultivated, harvested, cooked and eaten, perhaps by joining in these activities yourself. Each way of gathering data complements the others to give a holistic idea of how humans interact with the botanical world around them.

The precise methods of data collection you employ will change through the course of the study. You will likely begin a project by just living within a community for a while, taking the time to observe general characteristics of the local culture.

After getting to know people, you start to have exploratory interviews during which you attain familiarity with basic concepts and categories. After you have selected an issue for in-depth study, you can apply formal methods that allow you greater precision in selecting whom you interview and what you ask them.

Ethnobotanists working within their own culture can modify the methods according to their own needs. Keep in mind that recording oral traditions and applying other methods discussed in this manual may be unusual in your community, requiring you to take on the perspective of an outside observer of your own culture. It is often appropriate to seize this opportunity of acting as an outsider to gain a different view on local beliefs and customs, particularly when you cross generational, gender or class lines in your community.

4.1.1 Selecting local counterparts

The local people who share their cultural and ecological knowledge are called **informants** by most anthropologists. In some countries and social contexts, people find this term derogatory and prefer to use alternatives such as interviewees, subjects, participants, respondents, collaborators or local counterparts.

Whom do you ask about plants, animals and other elements of the natural environment? Stefano Varese, a Peruvian anthropologist, recalls his first visit to an indigenous village in the mountains of Oaxaca, sufficiently remote that he had to arrive by small plane. Soon after he descended from the aircraft, he was greeted by a villager who asked, 'Are you an anthropologist?' When Stefano responded yes, the eager villager said, 'Well, I'm an informant.' Although such a person might be able to provide a wealth of information, he is unlikely to be a typical villager.

Many ethnobotanists are casual in their choice of local counterparts. They speak to the first person encountered, ask for the individual who knows the most about plants in the community or interview anyone who is willing to talk with them.

Although it is acceptable to talk with the first people you meet when starting a project, many anthropologists recommend that informants be selected in a systematic way as the study progresses. If you are planning on using statistics to interpret your data, it is best to select a **random sample** that embodies a representative cross-section of the community. Choosing your informants randomly implies that all members of the population have an equal statistical chance of being included, ensuring that your sample will not be **biased** in favor of any particular social group (for example, more men than women or more old people than young people). This is particularly important in situations where the dominant social group – men, elders, wealthy or highly educated people – tends to be the first to come forward to be interviewed. Picking a random sample, or at least one that represents the diversity of the community, will ensure that you speak with the silent majority which often includes women and children, the poor and other groups who have often been ignored by fieldworkers.

Table 4.1 Some sociological variables used to describe the local participants in an anthropological study

Age
Ethnicity
Religion
Place of birth
Gender
Occupation
Migration to other regions for work or marriage
Age at marriage
Kinship and marriage relations
Number of children
Number of people in household
Number of generations in household
Literacy
Education
Language ability

There are many different ways of choosing a random sample. From a list of people or a map of houses, you could select every second (or third, or fourth, or fifth) person or house. You could also assign numbers to the people or houses, prepare slips of paper on which each identification number is written and then pick out as many slips as people or houses you wish to visit. The same result can be achieved by using a **random numbers table**, which allows you to choose numbers in an arbitrary way. Alternatively, you can flip a coin for every person or house, heads meaning that you will interview, tails meaning that you will not. These four approaches are illustrated in Box 4.1.

Choosing a random sample of informants is feasible when you are recording the general knowledge about plants in a community. In other cases, the choice of informants will be dictated by the subject of the study, such as when you are recording specialized knowledge held by traditional healers or by certain social groups such as children, women or the elderly.

Whether or not you choose your sample in a random way, you should characterize the local people who work with you and show how they represent the overall population of the community. Some of the different ways that a sample can be characterized and compared with a broader population are summarized in Table 4.1. These characteristics, called **sociological variables**, can be used to analyze the way people behave and the beliefs they hold. After noting the informant's name and residence (address or community), you should record gender (whether the person is male or female) and place and date of birth, from which his or her age can later be calculated.

Several questions may be asked about education and literacy: 'Is the person able to read and write?'; 'How many languages does he or she speak?'; 'How many years

Box 4.1 Choosing a random sample of households or people

Genstown is an imaginary village that comprises 30 households (Figure 4.2). Some people speak French, some speak English and some households are bilingual. French-speakers tend to be Catholic and live in the northern part of the village, while most English-speakers are Protestant and live in the southern part of the community (Table 4.2).

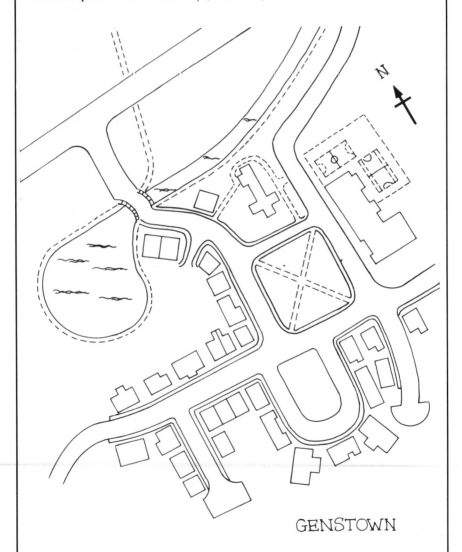

Figure 4.2 Map of houses in the imaginary village called Genstown.

Table 4.2 A partial household census of Genstown, including the house number, family name, language and religion of all of the resident families

House number	Family	Language	Religion	Skip house	Coin toss	Paper slips	Random number
01	Debucourt	French	Catholic	Yes	Yes	No	Yes
02	Jardin	French	Catholic	No	Yes	Yes	Yes
03	Lapierre	French	Catholic	Yes	No	Yes	No
04	Gereau	French	Catholic	No	No	Yes	No
05	Chevallier	French	Catholic	Yes	Yes	No	No
06	Moret	French	Catholic	No	Yes	Yes	No
07	Foliguet	French	Catholic	Yes	No	No	No
08	Sureau	French	Protestant	No	No	Yes	No
09	Nourissier	French	Protestant	Yes	No	Yes	Yes
10	Dupont	French	Protestant	No	Yes	No	No
11	Moreau	Bilingual	Protestant	Yes	Yes	No	Yes
12	Corbin	Bilingual	Catholic	No	Yes	No	No
13	Hailey	Bilingual	Catholic	Yes	Yes	Yes	Yes
14	Ludlum	Bilingual	Protestant	No	No	Yes	Yes
15	Collins	Bilingual	Protestant	Yes	Yes	Yes	No
16	Robinson	English	Catholic	No	No	No	Yes
17	Miller	English	Catholic	Yes	Yes	Yes	No
18	Anthony	English	Protestant	No	No	No	Yes
19	Smith	English	Protestant	Yes	Yes	No	No
20	Colbert	English	Protestant	No	Yes	No	Yes
21	Wolfe	English	Protestant	Yes	Yes	No	Yes
22	Hermann	English	Protestant	No	No	No	Yes
23	Wells	English	Protestant	Yes	No	No	No
24	Harden	English	Protestant	No	No	No	Yes
25	Ross	English	Protestant	Yes	Yes	Yes	Yes
26	Davis	English	Protestant	No	No	Yes	Yes
27	Washington	English	Protestant	Yes	Yes	Yes	No
28	Morris	English	Protestant	No	No	Yes	Yes
29	Jones	English	Protestant	Yes	No	Yes	No
30	Harrison	English	Protestant	No	No	No	No

Imagine that you arrive at the village and want to interview half of the households about what ornamental flowers they bring to church, but you do not know the language, religion or other information about the residents. You could choose a random sample of informants by: (1) visiting every other house, going from the northwest to the southeast corner; (2) flipping a coin in front of every house and interviewing the inhabitants when the coin comes up heads; (3) numbering each house on a map and then putting the house numbers on slips of paper, shuffling them in a hat and picking 15 slips; or (4) selecting 15 houses according to a random numbers table, such as the one shown in Figure 4.3.

7797	3118	4201	0834	0983	5549	0072	7890	3532	2814	9289	7268
1812	**1142**	5613	0275	0169	4405	7658	5157	3879	8481	6746	**2199**
6271	4550	7845	8852	6899	9639	7844	3809	4759	7248	4232	**0953**
0941	9873	3438	8844	1456	8056	6827	4855	4104	6550	8971	5088
1484	8553	**1314**	0607	5870	5826	6943	8700	2543	8090	9797	8890
3526	**2507**	7409	0367	7086	7169	8315	0231	6190	1218	**1674**	**2890**
6260	**2651**	5009	3912	8249	1635	0524	2139	1739	0921	8545	8035
3292	3781	**2253**	7832	0273	0627	4029	1649	3237	4784	7863	3259
7440	**0863**	4111	4556	3050	6268	8231	8491	9138	5099	4507	8904
9186	2685	8752	4615	4282	1961	2378	8874	2478	3334	3459	9980
2471	2449	5402	7058	7201	4572	3437	2068	5257	9686	6632	5551
7350	9404	2140	4927	3497	6423	7325	8370	2288	7846	4258	2827
4457	**0276**	**0183**	4604	9967	7770	6833	6959	8239	1933	3965	7143

Figure 4.3 In using this random numbers table to select houses from Genstown, I started with the first two digits in the 11th row, and then continued until I found 15 numbers between 1 and 30. I did not count duplicates of the same number, 00 or numbers 31 and above. The selected numbers are indicated in **bold.**

I followed each of these steps and obtained the results shown in Table 4.2. By picking every other house, by choosing half of the slips of paper or by selecting the house numbers from a random numbers table, I was guaranteed to come up with 15 houses. On average, 15 houses would be selected by a coin toss, although this technique could give you a few houses more or less (by chance I came up with exactly 15 heads and 15 tails when I carried out this experiment).

If we compare the resulting samples, we find that a representative cross-section of the community has been chosen (Table 4.3), even though a different assortment of houses was selected by each technique. This would not have been the case if we had used a non-random or biased method, such as interviewing only people in the northeastern sector of town or only those who come out of the Catholic church after Sunday mass.

Table 4.3 Comparison of informant groups obtained by using four different random sampling techniques

	1. Skip houses	2. Coin toss	3. Paper slips	4. Random number
French Catholic	4	4	4	2
French Protestant	1	1	2	2
Bilingual Protestant	2	2	2	2
Bilingual Catholic	1	2	1	1
English Catholic	1	1	1	1
English Protestant	6	5	5	7
Total	15	15	15	15

of formal education were completed?' Enquire about the informant's work, including both subsistence and wage labor. If he or she has migrated, ask for what purpose and the length of absence from the community.

Other queries may be pertinent, depending on the characteristics of the community in which the work is being carried out and the focus of the research. 'What is the person's ethnicity and religion?', 'When married, does a couple reside with the husband's parents (**patrilocal residence**), with the wife's parents (**matrilocal residence**) or do they set up a new household (**neolocal residence**)?', 'Does the typical household contain just parents and children (a **nuclear family**) or does it include grandparents, cousins or other relatives (an **extended family**)?'.

Whenever possible, the above dimensions should be related to indigenous categories. For example, American anthropologist Carole Browner recorded the age of people who responded to a survey in a Chinantec community of Oaxaca, Mexico. When analyzing the results, she split the interviewed population into three classes used in the community to decide who is eligible for public office: young adults aged 18–35 years, middle-aged adults from 36–45 years old and retired adults 46 years old and over. She then compared the responses of individuals from these three groups to an interview about the plants used for controlling fertility, easing birth and other aspects of human reproduction.

After recording information about the people you have talked with, you can demonstrate how this sample compares to the overall population of the community. If available, you can refer to a **census**, a complete list of all residents which includes information on their age, gender, occupation and many other sociological variables.

As one of their first activities in a community, many anthropologists make a census. They begin by drawing a map which shows the location of all houses. After assigning a number to each house, they visit all residents to request the information described above. When possible, these data are confirmed by consulting the birth and death records of the community and other secondary sources. The number and location of the houses and the sociological variables of each household member can be included in a database which facilitates analysis of the data.

If you do not have enough time to carry out a census, you can rely on secondary information which has been collected by other people. Some villages have carried out their own censuses or have copies of surveys made by anthropologists, governmental agencies or non-profit groups. National censuses often provide information that has been extensively analyzed and is presented on community, regional and national scales.

Be aware that national censuses are not always carried out with the same care and precision as those made by researchers and the communities themselves. National surveys are often completed in a short period of time by people who are unfamiliar with the community. Although overall population figures may be roughly accurate, there is always the possibility the census-takers will overlook people living in remote ranches or those who have nomadic lifestyles. Data on literacy rates, bilingualism and income are often unreliable, because these are issues that can invoke feelings of pride and jealousy. In countries in which fluency in an indigenous language is considered a sign of being culturally backward, people may not admit to being bilingual. Among ethnic groups in which accumulation of wealth is discouraged, household members may under-estimate their personal wealth. Some people, embarrassed that they cannot read or write, will erroneously state that they are literate. These modifications of the truth are even more common when the interviewer is a stranger who has not won the trust of local people and has no intuition about whether the responses are accurate or not. For these reasons, you should be sceptical about the information in national censuses and attempt to verify all data with community members.

4.1.2 Establishing rapport

I was once told a story, fictional I hope, about some botanists who traveled through Mexico and Central America to make germplasm collections of squashes. They would take notes and photographs of the fruits and then would break them open to harvest the seeds. At one point in their trip, they stopped to talk with a farmer who was harvesting his squash crop. After a few minutes of conversation, he selected some of his finest fruits and proudly gave them to the visitors. They thanked him, took some photographs of the fruits, smashed them on the ground in front of the farmer and scooped the seeds into a numbered bag which they threw into a box containing many similar bags. It was standard scientific procedure for the botanists, who did not stop to think that it was an inappropriate way to express their appreciation to the farmer.

Before you begin speaking with anyone, think about the impact that your non-verbal behavior has on people. Because many aspects of courtesy and good behavior are universal, common sense can guide your actions even when you have a limited understanding of local customs. In your own culture, would you accept a gift and then smash it to pieces on the ground?

Frank Lipp, an anthropologist from the United States, has given some useful advice to people who are about to begin research in a community [59]. Successful fieldwork, he writes:

> ... is dependent upon gaining entrée and establishing rapport with an indigenous group. To approach a local person with notebook and pencil and arbitrarily demand answers is the surest way to arouse resentment and reticence .... In order to create an atmosphere of trust, the field worker must exhibit a genuine sense of warmth, empathy and respect for his informants not by acting so but by being so in his actions and his words. The subject's ability to sense and respond to our warmth and/or coldness, our involvement or our facade, necessitates treating someone of another culture, ethnic group or world view as a respected equal and understanding him (or her) in terms

of his ideas and values rather than in terms of our own. Although the field worker remains committed to the values of his empirical discipline, he should behave as a human being, not as a technician, and cast aside any mask or 'professional role' that may create barriers between informant and researcher.

Your acceptance by the community will depend to a great extent on how you interact with others. You find that each element of behavior – your manner of dress, facial expressions, manner of speaking, openness to trying unusual foods and participating in new experiences – can endear people to you or make them feel estranged. The best advice is to be willing to learn about the correct way of behaving in another culture without forgetting the good manners of your own.

4.1.3 Reliability of field data

Anthropologists used to assume that by immersing themselves in another culture, they could become one of the local people, opening the way to a complete and loyal understanding of how others live. There is currently much scepticism about this belief. Many researchers now accept that no matter where they go and how long they live in a foreign land, they still retain **cultural filters** – their own personal ways of looking at the world, conditioned by how they were raised and educated.

Although it is impossible to become an objective observer, there are ways to diminish the impact of your subjectivity during fieldwork. We bring to the field many attitudes and habits which must be modified if we are to learn from other people. We must re-educate ourselves to become good listeners and to speak with people in a way that allows them to freely express their ideas and opinions.

Avoid enquiries that are aimed at eliciting specific responses that reinforce your previous observations or conclusions. Emphasize interactions in which local counterparts have the opportunity to express themselves in their own words without coercion.

Without being aware of it, we often pressure people into giving preconceived answers rather than allowing them to explain their own story. In open-ended interviews, anthropologists try to avoid directed questions, which encourage people to give responses that do not entirely reflect their true beliefs. For example, it is better to pose questions such as, 'Does this plant have a name?' rather than 'What is the name of this plant?' or 'Is the name of this plant *begonia*?'. The first question allows the respondents to say that they do not know or that there is no name. The other questions demand a response and encourage people to improvise or agree with an answer.

In everyday conversation, we sometimes pose questions in a way that limits people to one of two answers – 'yes' or 'no', 'right' or 'wrong', 'here' or 'there'. Anthropologists, when beginning fieldwork, refrain from asking dichotomous questions, because the truth might lie somewhere in between. Posing questions in an open way allows informants to rephrase the enquiry, making it more relevant to their own culture and knowledge. Instead of asking for a discrete response it is better to stimulate people to give a detailed explanation in their own words.

In order to become a good interviewer, you must first become a good listener. What might be considered normal in some contexts – interrupting people with comments or questions, helping them to answer or rewording their responses – is likely to disturb the flow and content of the interview.

The rhythm, length and content of conversations may vary from one culture to another. Make the effort to adapt yourself to the local style of interaction. Do not be in a hurry when talking with people but be aware when they are growing impatient to get back to their own work. Choose the time of the interview according to local schedules and not your own. Although your status as an outsider may allow you to pose questions that would be inappropriate if raised by a community member, be tactful when discussing situations which may be delicate, such as personal wealth or religious beliefs.

Even if you follow these suggestions, cultural bias inevitably slips into your interactions and conversations with community members. Researchers and local people, when reviewing the results of your fieldwork, may wish to examine the raw data, including how questions were posed and how interviewees responded. Whenever possible, you should tape your interviews with a cassette recorder. Alternatively, you can record the questions and answers word-for-word, leaving your own interpretation or rewording aside. Some anthropologists record visual images of their interactions with people, using either a still or video camera.

Another way of decreasing the impact of your own cultural bias is to turn the tools over to the community. With video cameras, notebooks and tape recorders in hand, people can record their own perceptions of local beliefs and customs, providing a valuable record for the community archives as well as for visiting researchers. In Comaltepec, for example, I asked a well-known Chinantec curer named Roberto Hernández López to write about his experiences. The resulting manuscript, written in Spanish, describes how he became a curer, how he diagnoses patients and how he treats various health problems. It includes a list of plants and the illnesses they treat, all called by their Chinantec names.

Even when ethnographers take care in posing questions and recording interactions, it is probable some erroneous information will be given intentionally by a few informants. In almost every community, there are a few people who give fictitious accounts to unsuspecting outsiders or who contradict what other people have said. This may be an attempt to put nosy anthropologists in their place or it may simply be a way of having fun. There are others who seek to gain prestige by exaggerating their own knowledge, making up data in the process. If compensation is offered in an inappropriate way, some informants will be tempted to embellish their stories to establish themselves as a good source of information.

Detecting erroneous data is an ability that comes with experience. After spending some time in a community, you begin to have a feeling for which answers are valid and which people are sincere. As you become increasingly competent in the local dialect and aware of indigenous classifications, your questions will be more and more precise, encouraging people to discuss their knowledge in an accurate and elaborate way. Much insight comes from **cross-verification**, which is checking to see if what one person tells you is consistent with the versions of other informants and if what you learn from one approach reinforces what you discover with another. This process of validation is referred to by some researchers as **triangulation**, which comes from the idea that observing an object from diverse perspectives gives you the best view of all.

The best way to collect valid data is to create a good relationship of cooperation and a sense of mutual benefit in the communities where you work. When the goal is to improve local health care, safeguard natural resources, strengthen cultural knowledge or provide alternative sources of income, the few people tempted to improvise data will be quickly discouraged by their fellow villagers.

4.1.4 Keeping a secret

An important part of building cooperation is respecting local people's desire for confidentiality. At times, people will tell you something only on condition that you do not identify them when you talk to others or in written accounts of your work. Some communities which prefer to stay anonymous will request that you use a pseudonym instead of the village name in all official reports. Some individuals, fearful of political repression, will request anonymity.

In other cases, you will be given information that people do not want disclosed at any time or in any form. Jorge Luis Borges, an Argentinian author, wrote a short story about a young ethnologist in such a predicament. He is sent by his professor to learn the spiritual secrets of a North American Indian group as part of the work towards his doctoral thesis. After many months of patient observation and participation, he is initiated into a religious society and told the secrets of the tribe, under the condition that he never reveal anything to non-initiates. When he returns to the university he has a choice to make – publish his findings or give up his career as an anthropologist. He chooses to keep his promise to the Native Americans, much to the dismay of his professor.

Although this is an extreme case, you are likely to face similar decisions about the knowledge you gain from certain sources. Yíldiz Aumeeruddy, an ethnoecologist from Mauritius, was told two myths by rural Indonesians that explain restrictions on land use in Jujun, a village near Kerinci Seblat National Park in Sumatra [60, 61]. Although one legend was common to many villages in the region, the other was only known in Jujun. People feared that disclosing it to other people would create new land conflicts in the area. Although she uses the first myth in her analysis of Kerinci ethnoecology, she has decided not to reveal the second legend out of respect for this taboo.

4.1.5 Participant observation and keeping a field diary

Participant observation is a technique we all use as soon as we set foot in a community which is not our own. To overcome our feeling of **alienation** – that many things around us are strange – we begin to take stock of what is different and what is similar to our own culture, including language, eating habits and many other elements of everyday reality. We often act as children in this new context, trying to manage with a limited vocabulary, faulty grammar and limited understanding of what others say to us. There are new motor skills to learn. We are invariably clumsy when we first start using unfamiliar agricultural implements, eating with unusual utensils and playing new games.

Claude Levi-Strauss, a French anthropologist, provides several perspectives on the initiation of participant observation in his book *The Savage Mind*, which is about symbolic classification of the natural world [62]. He quotes an amusing account written by Elizabeth Smith Bowen, a British anthropologist, of her first encounter with local language and ecological knowledge in Africa [63]. It gives an accurate feeling of the first bewildering days that fieldworkers spend in a foreign land, trying to compare another culture with their own:

> These people are farmers: to them plants are as important and familiar as people. I'd never been on a farm and am not even sure which are begonias, dahlias or petunias. Plants, like algebra, have a habit of looking alike and being different, or looking different and being alike; consequently mathematics and botany confuse me. For the first time in my life I found myself in a community where ten-year-old children weren't my mathematical superiors. I also found myself in a place where every plant, wild or cultivated, had a name and a use, and where every man, woman and child knew literally hundreds of plants ... (my instructor) simply could not realize that it was not the words but the plants which baffled me.

Although this stage of field research can be frustrating, it is rich in discovery. Most anthropologists attempt to record as many observations as possible in a field diary, which includes comments on relevant events of each day. For example, if your object is to record knowledge about plants, you note the local names and uses of every species that people tell you about, that you observe being used or that you help to gather and prepare. At this stage of the research, you probably will not be able to identify all the plants or transcribe the local names, but you can review the field journal at a later date to fill in this information. Box 4.2 contains a passage that Claude Levi-Strauss quotes from the detailed field journal of Harold Conklin, an experienced fieldworker who has been studying ecological knowledge of indigenous people in the Philippines for over 30 years.

There is no strict method to guide participant observation. The most important tools are curiosity, a willingness to learn from other people and an ability to adapt to their rhythm and lifestyle. Although there is a popular idea that anthropologists

Box 4.2 Keeping an ethnobotanical journal

Harold C. Conklin, a professor of anthropology at Yale University in the United States, wrote a doctoral dissertation in 1954 entitled *The Relation of Hanunóo Culture to the Plant World*, which describes the ethnobotanical knowledge of an indigenous group in the Philippines [64]. Although never published, it is considered one of the classic works of ethnoscience because it introduced an empirical approach to studying local knowledge of the natural environment. Early in the manuscript, Conklin gives an excerpt from his field diary. After recording these notes in the field, he annotated the text with scientific names and bibliographic references when he returned home. As Claude Levi-Strauss quotes:

At 0600 and in a light rain, Langba and I left Parina for Binli .... At Aresaas, Langba told me to cut off several 10 × 50 cm strips of bark from an **anapla kilala** tree [*Albizia procera* (Roxb.) Benth.] for protection against the leeches. By periodically rubbing the cambium side of the strips of saponaceous (and poisonous: Quisumbling, 1947, 148) bark over our ankles and legs – already wet from the rain-soaked vegetation – we produced a most effective leech-repellent lather of pink suds. At one spot along the trail near Aypud, Langba stopped suddenly, jabbed his walking stick sharply in the side of the trail and pulled up a small weed, **tawag kugum buladlad** [*Buchnera urticifolia* R. Br.] which he told me he will use as a lure ... for a spring-spear boar trap. A few minutes later, and we were going at a good pace, he stopped in a similar manner to dig up a small terrestrial orchid (hardly noticeable beneath the other foliage) known as **líyamlíyam** [*Epipogium roseum* (D. Don) Lindl.]. This herb is useful in the magical control of insect pests which destroy cultivated plants. At Binli, Langba was careful not to damage those herbs when searching through the contents of his palm leaf shoulder bag for **apug** *slaked lime* and **tabaku** [*Nicotiana tabacum* L.] to offer in exchange for other betel ingredients with the Binli folk. After an evaluative discussion about the local forms of betel pepper [*Piper betle* L.] Langba got permission to cut sweet potato [*Ipomoea batatas* (L.) Poir.] vines of two vegetatively distinguishable types, **kamuti inaswang** and **kamuti lupaw** .... In the camote patch, we cut twenty-five vine-tip sections (about 75 cm long) of each variety and carefully wrapped them in the broad fresh leaves of the cultivated **saging saba** [*Musa sapientum compressa* Blco. Teoforo] so that they would remain moist until we reached Langba's place. Along the way we munched on a few stems of **tubu minuma**, a type of sugar cane [*Saccharum officinarum* L.], stopped once to gather fallen **bunga**

> **area** nuts [*Areca catechu* L.] and another time to pick and eat the wild cherry-like fruits from some **bugnay** shrubs [*Antidesma brunius* (L.) Spreng.]. We arrived at the Mararim by mid-afternoon having spent much of our time on the trail discussing changes in the surrounding vegetation in the last few decades!

are forever observing and participating in dances, festivals, marriages, funerals and other ritual occasions, most of their work involves the everyday activities of local people. Be prepared to accompany people as they bring in the harvest, collect plants in the forest, build houses and tend to their animals. The longer you stay, the more complete your experience will be. Many anthropologists try to live in a community for at least a year, allowing them to observe the changes that come with each season.

In the context of these everyday activities, ritual events will give you a special insight into aspects of the culture that are not evident at first sight. Alejandro de Avila, a Mexican researcher, had spent much time in Coicoyán, a Mixtec village in western Oaxaca, before he was able to participate in the ritual of San Marcos, celebrated every year on April 25th [65]. He accompanied a group of residents to the top of a local hill, where they burned incense, lit beeswax candles and sacrificed chickens near the shrine of San Marcos. An elderly man whispered prayers and left specially prepared bundles of rushes around the shrine. This and other special occasions allowed Alejandro de Avila to understand the use of several ritual and hallucinogenic plants, currently used less and less by the Mixtec people, which he had not seen as he observed everyday life in the village.

4.1.6 Open-ended conversations

Intuition and experience are the best guides to informal ways of gathering information. When beginning fieldwork, we are drawn into a broad range of conversations. With inspiration and good luck, we find ourselves asking the questions that open the way to understanding a foreign culture. Although these initial dialogues will cover some issues not clearly linked to ethnobotany, you will find that observations about local agriculture, medicinal herbs, hunting and other relevant subjects will naturally arise. Through these first open-ended interviews, you will develop a sense of what needs to be asked in more structured interviews.

As people begin to understand what interests you, discussions are likely to drift more and more towards ethnobotany. As you walk in the forest, work in agricultural fields or go hunting with local people, they naturally talk about their perceptions of the environment around them. When you visit people in their home, they show you unusual medicinal herbs in their home gardens, artifacts made from natural materials or fruits brought back from the forest, all of which spark conversations about local botanical knowledge.

Even in the absence of these props, you can easily guide the conversation towards ethnobotany because plants and animals have an important role in all subsistence and ritual activities and are a natural focus of attention. One option is to bring your own plants with you, either fresh or dried. Many researchers show pressed specimens when interviewing informants. Even though the plants are removed from their ecological context and are missing some of their morphological features, most can be recognized without difficulty.

You can take walks through home gardens, agricultural fields or forested areas to enquire about plants that attract your attention or that of your local counterpart. Some fieldworkers plan **forest or garden transects** in advance, ensuring that they pass by plants and vegetation types of particular interest. James Boster, an anthropologist from the United States, planted dozens of local varieties of manioc (*Manihot esculenta* Crantz) in a garden near an Aguaruna community in the Peruvian Amazon [66, 67]. After the plants grew to maturity, he led informants through the garden, asking for the name and the main morphological characteristics of each cultivar. From this exercise, he was able to learn that women, the cultivators in the community, know much more about manioc than men and that knowledge about the cultivars varies significantly from one family group to another.

Another approach is to choose a particular **ethnobiological artifact**, an object made from plants or animals, and discuss all of the species associated with its manufacture and use. Sandra Banack, an American ethnobiologist, used this tactic when studying boat-building in Polynesia [68]. Although the making of outrigger canoes is a rare event these days, she encouraged a group of Fijian men to reconstruct the various steps. She identified the wood used for the main body and the outrigger and the fibers used for the mast. She documented the special foods eaten by the boatmakers, which were prescribed by **cultural taboos**, or restrictions on diet and behavior. Finally, she recorded how foods were prepared for sea voyages. In this way, starting with a single artifact, she discovered the use and name of dozens of plant species.

4.1.7 Semi-structured interviews

All these approaches will lead you to **semi-structured interviews**, in which some questions are determined beforehand and others arise during the course of the conversation. Before you begin an interview, prepare a checklist of topics and questions that you would like to cover. As the discussion gets under way, new lines of enquiry will arise naturally and you will let some of your prepared questions fall to the wayside, left for a future discussion. The inspiration for the questions will come from your previous interactions in the community, including participation and observation of everyday activities as well as informal conversations guided by artifacts, plant specimens or transects. This background will allow you to ask culturally appropriate questions, understand answers and improvise follow-up enquiries.

At some point in the interview, ask your partner for his or her age, name and other personal information which will help in the interpretation of the responses. Although you may wish to request these data at the beginning to ensure that you record the same variables for each person, it is often less imposing to weave these questions into the flow of the discussion.

Keep in mind that the dynamics of interviews are determined not only by the respondent but also by the interviewer. Differences in age, gender, social position, ethnicity or religion affect the quality of the interaction. Depending on the topic, women interviewers may get more personal information from other women, whereas men may prefer talking to other men about certain issues. Young inter-viewers may have difficulties eliciting information from older, more prestigious people. When possible, interviews should be carried out by a team of co-workers who are culturally diverse. These interviewers can go together to the selected households or they can visit sequentially over a period of days, thus achieving well-rounded information.

Personal rapport is affected not only by the social standing of the interviewer but also by his or her attitude towards the respondent. Attempt to sit on the same level and in the same position as community members. Begin and end the conversation with polite talk, if this is the local custom, and respect other norms of socializing that you observe in the region. Whenever possible, adopt local ways of speaking, using expressions and words that people can readily understand. Learn to understand gestures used locally and to read other body language of your local counterparts.

Most semi-structured interviews are conducted with a single person at a time. This allows people to express personal viewpoints, discuss disagreements in the community and speak freely without being interrupted or contradicted by others, as often happens in group interactions. Sensitive information concerning personal income or health conditions is more likely to come out in the intimacy of individual interviews than when many people are around.

Group interviews also have their place in anthropological fieldwork. Partici-pants can build consensus by discussing an issue amongst themselves and agreeing on an answer – or agreeing to disagree. This allows a broad view of the scope of opinions in a community and possible points of contradiction. Be aware of the dynamics of the group interaction and seek individual interviews with people who do not have a chance to express their opinion during the meeting.

Although the group interview may be held with a randomly-assembled group of people, you may wish to arrange some meetings between specialists who are particularly knowledgeable about a certain topic. These **focus interviews** can build upon previous meetings with key informants whom you have identified during the course of the fieldwork, creating a unique opportunity for specialists to share expertise and discuss their knowledge in greater depth.

Posing questions, listening to answers, reading body language and studying the dynamics of the interaction will keep you busy during the course of the dialogue. Keep in mind that, even while doing all of these things, you must also record as literally as possible people's responses and your own impressions. Although many researchers tape all of their interviews, there are disadvantages to this approach. Some local people may be reticent to express freely their opinions when they are being recorded. And taping requires double work, because each cassette will probably take twice as long to transcribe as it did to record. The traditional alternative is to write all questions and answers in a field notebook. Some researchers divide each page into two sections, one side for the literal answers of informants and the other to record interpretations and observations. At the end of the day, review your notes to see if you have any additional observations to record.

No matter which technique is chosen, you should explain to your local counterpart how and why you are recording the conversation and what will be done with the results. Although you should show discretion by using a small notebook or recorder, never attempt to hide the fact that you are recording the dialogue. Leave copies of your notes or cassette tapes with the people you inter-view, if they so request.

4.2 Searching for ethnobotanical information in folklore

Although conversations, interviews and participant observation will be the source of most of what you learn in the field, be aware of the unique perspective provided by **myths and legends** as well as songs, ritual incantations and sayings. Myths and other traditional stories recount often fantastical historical events which are said to explain natural phenomena, such as the birth of the sun and the moon or the origin of cultivated plants, as well as cultural practices, including dietary restric-tions, hunting taboos and agricultural rituals.

The French anthropologist Claude Levi-Strauss, who has written extensively on how to analyze myths, suggests gathering as many versions as possible and then comparing them to discover which events and elements are common to all. These recurring features reveal the underlying structure and meaning of the myth. Levi-Strauss observed that there is often an interaction between the natural and the human world in myths and that a symbolic parallel is drawn between the two. Box 4.3 explores the symbolic and empirical aspects of myths concerning the origin of *kava*, a psychoactive plant used in Polynesia.

Box 4.3 Polynesian origin myths of *Piper methysticum* Forst. f.

Piper methysticum is a psychoactive plant which has been cultivated in many parts of the Pacific, from Papua New Guinea eastward to Tahiti. Known by the Polynesian name **kava** and by numerous other local names, it is the source of a drink that is consumed by Melanesians, Micronesians and Polynesians during public ceremonies and private gatherings. Botanists link

---

its origin to a wild species, *P. wichmanni*, which is endemic to New Guinea, the Solomon Islands and Vanuatu. Long before the arrival of Europeans to the area, local people brought this species into cultivation. *Piper methysticum* is now completely domesticated. It never produces seeds and is entirely dependent on vegetative reproduction, carried out by cultivators, for its survival.

Vincent Lebot, a French botanist who has studied in depth the phyto-chemistry, morphology and cultural use of **kava**, recounts several myths about its origin that he recorded during his fieldwork or found in ethnographies [69]. He gives examples of related myths from four different islands.

A very long time ago, orphan twins, a brother and a sister, lived happily on Maewo. One night, the boy, who loved his sister very much, had to protect her from a stranger who asked to marry her but whom she had refused. In the struggle, the frustrated suitor loosed an arrow which struck the boy's sister and killed her. In despair, the brother brought his sister's body home, dug her a grave and buried her. After a week, before any weeds had grown over her tomb, there appeared a plant of unusual appearance which he had never seen. It had risen alone on the grave. He decided not to pull it up. A year passed and the sorrowful boy had still not been able to quell the suffering he felt at his sister's death. Often he went to mourn by her grave. One day, he saw a rat gnaw at the plant's roots and die. His immediate impulse was to end his own life by eating large amounts of these roots, but instead of dying he forgot all his unhappiness. So he came back often to eat the magic root and taught its use to others. [Maewo island in Vanuatu]

Kava first came to Samoans through Tagaloa, the first **Matai** or chief. Tagaloa had two sons, Ava and Sa'a. As Ava lay dying, he murmured to Sa'a that from his grave would come a plant of great value to the Samoan people. Ava died and was buried. Sa'a and his children watched the grave and on the third day after Ava's burial, two plants were seen growing from the head of his grave. As Sa'a and the children watched, a rat came and ate the first plant. It then moved to the second one and began to eat, but quickly became intoxicated. The rat went staggering home as the people watched in astonishment. They named the first plant **Tolo** or sugarcane and the second **Ava** in honor of the man from whom it sprung. [Upolu Island in Western Samoa]

On the Island of Euaiki, the chief Loau recognized human flesh at a meal and told people not to eat it – it should be planted in the ground and brought to him when it matured into a plant ... the body grew up into a **kava** plant arising from different parts of the body. And when it matured he noticed that a rat chewed on the **kava** and became paralyzed. [Tonga]

In one legend, the discovery of **sakau** is attributed to a rat nibbling a root and acting quite intoxicated. In another legend, more detailed, the origin is traced to Pohnpeian god Luk. The skin of a heel of a mortal man, Uitannar, was given by Luk to a woman in payment for her kindness. She was told to bury the skin and a plant would grow in its place. The juice of the plant would make people intoxicated and change their lives. This was done and **sakau** plant was later spread throughout Pohnpei. [Pohnpei]

Even though the stories are slightly different, they have several elements in common. They refer to not only symbolic interactions between humans and the natural world but also empirical methods used to discover the medicinal and psychoactive properties of plants. In each case, a human grave gives rise to a plant which is useful to people because it is intoxicating or has magic properties. Rats – wild animals that live near human communities – teach people how to use the plants.

This mixture of natural and cultural phenomena and the occurrence of species which are on the edge between wild and domesticated are common elements of myths which explain the origin of agricultural crops, medicinal herbs, domesticated animals and other organisms which play an important role in the lives of local people.

Figure 4.4 Line drawing of *Piper methysticum* Forst f., a domesticated beverage plant of Polynesia. In many local myths, the narcotic effect of the plant is said to have been discovered by watching the behavior of rats which either became drugged or died after eating the roots.

Following this comparative method, some researchers have sought to understand the medicinal and ritual uses of plants and animals by exploring in depth the symbolic interaction between culture and nature in myth. Plants and animals which are considered to be **anomalous** – abnormal in their behavior or morphology – often serve as symbolic **mediators**, known for their special ability to cure, nourish and protect people. These include domesticated species which, because they depend on people for their survival and have special morphological features which distinguish them from wild species, are thought to be on the interface of culture and nature. This is exemplified by the *kava* plant which, from a mythical perspective, arises from human flesh and is dependent on humans for survival, yet is similar to wild species which grow in the forest. Animals and plants which are **anthropogenic**, living near communities and in ecological niches created by human disturbance, are also considered as symbolic mediators. In the *kava* myths, rats – wild animals that forage in and around communities – show humans how to use a plant with medicinal and curative value. This relationship between wild and domestic things is just one of a series of opposites that symbolic anthropologists look for when they analyze cultural traditions. Many myths also draw contrasts between day and night, sky and earth, men and women, plants and animals and many other paired concepts that reveal a people's **cosmology**, or how they perceive the origin and order of the universe.

It is simpler to give examples of symbolic analysis than to explain how to carry it out. There is no set methodology. It is a matter of using intuition to gain insight into the meaning of another culture, which many symbolic anthropologists say they read and interpret as if it were a literary text. Nigel Barley, a British anthropologist who has written an amusing and perceptive account of his fieldwork among Dolawayo people of Cameroon, has spoken of his own struggles to analyze the symbols of another culture [70]:

> The problem of working in the area of symbolism lies in the difficulty of defining what is data for symbolic interpretation. One is seeking to describe what sort of a world the Dolawayo live in, how they structure it and interpret it. Since most of the data will be unconscious, this cannot simply be approached by asking about it. A Dolawayo, when faced with the question, 'What sort of a world do you live in?' is rather less able to answer than we ourselves would be. The question is simply too vague. One has to piece it all together bit by bit. Possibly a linguistic usage will be significant, a belief or the structure of a ritual. One then seeks to incorporate it all into some sort of a scheme.

In addition to the symbolic relationship between nature and culture, myths reveal many aspects of people's empirical knowledge of plants, animals and other elements of the natural environment. In the *kava* example, each myth made reference to people's observation of the rats which, after feeding on the roots of the plants, became intoxicated or died. Apart from the behavior of animals and the properties of plants, traditional stories reveal how people classify living organisms. For example, cultivated plants and their wild relatives may be considered as related pairs, indicating an appreciation of the relationship between similar species.

Most myths have a moral which focuses on the consequences of acting in a socially acceptable or unacceptable way. Good behavior, such as protecting relatives or burying the deceased in the proper way, brings rewards whereas poor behavior – breaking cultural taboos and deviating from social norms – is punished. Researchers have recorded many stories that describe the effects of not respecting traditional ways of caring for the natural environment. These legends often describe people who suffer misfortune because they have polluted a lake, cut a tree in a sacred grove or killed an animal without first asking permission from its soul. In this way, myth provides a code of behavior that allows people to get along with each other and to live in harmony with nature.

Some myths which describe correct social behavior have the secondary effect of encouraging people to preserve and manage natural resources. Yildíz Aumeeruddy has recorded an incest myth from the Indonesian villages of Jujun and Keluru, which are located near Kerinci Seblat National Park in Sumatra [60, 61]. It relates the story of a brother and sister who fall in love with each other through the misuse of a love potion made from bamboo oil. Although this myth is primarily about the taboo on incest, it has played an important role in preserving a managed forest that lies between the two villages. The sinning couple is said to have been thrown in a deep valley called *lura pungok* and this is the reason the villagers of Jujun give for restricting their activities in this forest to collecting plants and hunting wild pigs.

If you plan to collect myths as part of your research, try to collect versions from as many different people as possible. Once you have heard the story once, you can ask other people in the community or in the region if they could tell it to you in their own words. There are many details in a myth that may seem insignificant at first, but later serve as keys to understanding its underlying meaning. For this reason, be careful to write down the stories you hear with precision. Better still, tape them with a cassette recorder so that you can later transcribe them word for word. Although people will not always be able to tell you the moral of the story, it is worthwhile asking what each tale means. As you come up with an interpretation based on comparing numerous versions of the myth, ask local people what they think about your impressions.

4.3 Surveys and analytical tools

4.3.1 Structured interactions

The open-ended and semi-structured interviewing techniques discussed above are readily adapted for structured interactions, in which a selected group of people are requested to answer the same series of questions or to perform the same tests. These

Table 4.4 A model table showing how to construct a matrix that compares the responses of m informants on n plants or artifacts (the letters m and n are used to indicate indefinite numbers; in a real example the table would show the total number of informants and plants or artifacts involved in the study)

Respondent	Response on plants or artifacts				
	1	2	3	4	n
1	1, 1	1, 2	1, 3	1, 4	1, n
2	2, 1	2, 2	2, 3	2, 4	2, n
3	3, 1	3, 2	3, 3	3, 4	3, n
4	4, 1	4, 2	4, 3	4, 4	4, n
m	m, 1	m, 2	m, 3	m, 4	m, n

interactions yield comparative data which can be arranged in a matrix, a rectangular table in which the response of each informant to each inquiry is given. This matrix, modeled after the one shown in Table 4.4, can then be used as the basis for various statistical analyses.

There are two general forms of structured interactions. In **surveys**, selected informants give answers to a number of specific questions that are either written down in questionnaires or are posed in person by the interviewer. Surveys are the primary tools of sociologists, who seek to gauge the variation in responses between different groups of people in large and complex societies.

Through the use of various **analytical tools**, participants show their cultural preferences and empirical ecological knowledge by ranking objects or sorting them into piles based on criteria such as similarity or relative quality. Many of these techniques were developed in the 1960s by cognitive and linguistic anthropologists who were seeking systematic methods of collecting data which would reveal how people perceive and classify the world around them. Some of these approaches have been modified for use in participatory rural appraisal by simplifying both the methodology and the analysis of results.

Some fieldworkers shun surveys and tasks, because they consider them an impersonal way of collecting data. They argue that because they are based on brief interactions between the researcher and a number of respondents, it is difficult to establish the rapport that a fieldworker usually builds with local people during open conversation and participant observation.

Keep these criticisms in mind if you decide to carry out surveys and tasks as part of your research. These approaches are most effective when they are employed after you have become better acquainted with the people of a community. If you have limited time, put emphasis on carrying out the work in a participatory way. Design and carry out the surveys in collaboration with local counterparts with whom you can later analyze the results. If participants understand the goals and methods of the study and acquire skills they can use in community projects, these structured interactions become much more personal affairs.

4.3.2 The results of surveys and analytical tools

Four different types of results are obtained by surveys and analytical tools, exemplified by the following questions [58]. There are **inventories**: 'What are all native plant categories?'; 'What are the names of all the woody plants used for firewood?'. There are **rankings**: 'Among all fruits eaten, which are considered to be the best?'; 'Of all herbal remedies for headaches, which are the most effective?'. There are assessments of **similarities**: 'In an assortment of ornamental flowers, which are most alike?', 'Do people from one kin group classify plants in the same way, and differently from other people in the community?'. And there are **competencies**: 'Does everyone know the right name for a certain plant?'; 'Can all respondents accurately identify the curative properties of a series of medicinal plants?'.

The more structured an interaction, the more sophisticated are the statistical methods that we can use to analyze the results. But keep in mind that a highly structured interview yields data on a very limited part of local knowledge and takes much time and patience on the part of the researcher and local people. Sceptics of these empirical approaches feel that when a task removes people from familiar activities and habits, the results are likely to be an artifact of the method used. They say, for example, that when you ask someone to identify herbarium specimens, the responses will be affected by the fact that local people are not accustomed to looking at dried plants flattened in two dimensions and removed from their ecological and social context. In response to these doubts, verify if your results are consistent with the data collected through open interviews, participant observation, surveys and other tasks. Remember that this cross-verification allows you to look at local knowledge from several different perspectives, attaining a well-rounded view of how people interact with their environment.

4.3.3 Surveys

There are two types of surveys. A **questionnaire** is used when gathering comparable information from a broad range of literate informants, usually in an urban setting. The respondent fills in the questionnaire alone, which implies that there is little or no contact with the researcher. An **interview schedule** is a set of questions, usually written on a prepared form, that guides interviews with literate or non-literate individuals. These forms are used by the researcher in face-to-face interactions with informants, implying some degree of personal rapport. Both questionnaires and interview schedules are typically printed on paper, ensuring that the same questions are presented in the same way to all participants [71, 72].

One section of the survey is dedicated to recording the sociological characteristics of each individual, as discussed above in the section on selecting informants. The other parts are devoted to requesting the specific information that you wish to record. After the survey is completed, you may need to take additional steps to verify and further understand the responses. For example, if you request a list of medicinal plants, you will need to make herbarium vouchers of all the species mentioned. If informants give you the names of folk illnesses, you may need to hold separate interviews with local curers to document in more detail the symptoms and treatment of each.

The surveys of TRAMIL (discussed in Box 3.6), are prepared by a diverse group of colleagues – health care professionals, researchers and local people – who focus on herbal remedies used to treat the most important illnesses that affect a specific region. In some of the surveys, general data are collected on the informants, including their school attendance, living conditions and access to health services. The respondents are asked whether the chosen illnesses are treated at home, by a traditional curer or by a Western medical practitioner. For conditions treated with household herbal remedies, specific information is requested about the plants employed, preparation, parts used and other ethnobotanical data.

Lidia Giron, a Guatemalan pharmacist, carried out an ethnobotanical survey with other Guatemalan colleagues as part of her participation in TRAMIL [73]. With an interview schedule that followed the general TRAMIL format, they recorded verbal responses given by 403 people from four distinct ethnic groups in Livingston, Guatemala. Of these, 259 (64.3%) used plants for medicinal purposes, while 144 (35.7%) said that they did not employ herbal remedies. By studying the sociological data collected as part of the survey, Lidia and her colleagues discovered that the people most familiar with medicinal plants were middle-aged to older women, many of them housewives. Most of them could read and write Spanish and many still spoke *garifuna*, an indigenous language. The researchers took care to collect a fertile specimen of each plant mentioned by the respondents and they deposited these in an ethnobotanical herbarium in Guatemala City. This allowed them to identify quickly the species name of 102 of the 119 medicinal plants.

4.3.4 Survey questions

What kinds of questions are commonly posed in surveys? Unlike open-ended and semi-structured interviews, surveys usually involve queries that require short responses, often drawn from a set number of options. **Dichotomous questions**, those which can be answered 'yes or no', 'true or false' or 'agree or disagree', are common. These are a simple form of **multiple-choice questions**, in which the respondent is given a limited number of possible answers. For example, you might ask if a plant is used for 'medicine', 'food', 'condiment', 'firewood', 'construction' or 'ornament'. Whenever possible, it is preferable to draw the various possible responses from an **emic** perspective, that is, in a way that reflects local perception and classification of the natural world. In addition, you should give people the option to respond 'other', to indicate categories they feel are missing or to answer 'none of the above'.

If these questions are to be avoided in informal interviewing, why should they be acceptable in surveys? Most researchers agree that if questionnaires are to be used at all, they should be applied towards the end of a project. This leaves plenty of time at the beginning of fieldwork to experience the culture on its own terms and to learn by asking open-ended, spontaneous questions. Before a survey is attempted, the researcher should have an idea of the range of possible answers to a question. Dichotomous and multiple-choice questions provide a way of probing more deeply into matters that have already been discussed in free-format interactions. For example, you may pose the question 'Do you use chamomile tea as a medicine?' if you have discovered that there are people who do use the herb and others who do not and you wish to verify the relative frequency of use in the community. Including a query in your survey such as 'Which is the best source of firewood? [] oak, [] pine, [] mahogany, [] other_____' implies that you have already explored with people the range of woods that are available and which are preferred by local people.

Another important element of surveys is the **fill-in-the-blank** format, which consists of questions followed by a blank (_____) that the interviewer fills in with the answer given by the respondent. This type of query, which is used casually in open-ended and semi-structured interviews, includes all questions that begin with the six openers Who?, Why?, What?, Where?, When? and How? In surveys, these questions are written out and are posed systematically to a selected group of informants. Many questions familiar to ethnobotanists are fill-in-the-blank type, including: '*What* is the name of this plant?'; '*How* is it prepared?'; '*When* do these birds migrate through the community?'; '*Who* taught you about the medicinal properties of plants?'; '*Where* can this animal be found?' and '*Why* do you use this remedy instead of that one?'.

As with dichotomous and multiple-choice queries, fill-in-the-blank questions are posed only after you have gained some idea of local practices and knowledge and have established rapport with informants. Asking these questions during your first days in a community could lead people into improvising responses to please you or not to appear ignorant. Once you have established the ground rules of the research and local counterparts feel comfortable about sharing their knowledge, increasingly precise questions can be posed.

One of the main advantages of using short-answer queries is that they are much easier to analyze with statistics than the results of open-ended interviews. Conversational responses must be interpreted, categorized and coded before they are analyzed, which is time-consuming and creates the possibility of introducing researcher bias into the results. Survey answers, and the results of tasks described below, are **verbatim responses**. The exact words or opinion of the informant can be directly analyzed with only minimal polishing of the data.

Statisticians rate these various types of queries according to the related concepts of ease of guessing the answer and information content. For dichotomous questions, there is a 50–50 chance of guessing the right answer. This rate decreases somewhat with multiple-choice questions and even further with fill-in-the-blanks. Although some speculation is a normal part of any conversational answer, there is little probability than an open-ended response would be totally a result of guesswork. When people are given the opportunity to express themselves freely, they tend to focus on what they know.

The easier it is to guess an answer, the lower is the information content of the response. For this reason, fewer multiple-choice and fill-in-the blank queries are needed to obtain results that are statistically significant than is the case for dichotomous questions. When there is a high probability of guessing an answer, there is also a relatively high chance that agreement between informants is a matter of coincidence, which means that a large number of questions must be asked to prove that a result is statistically significant. An overview of the various types of questions employed in interviews and surveys is given in Table 4.5.

4.3.5 Analytical tools

Researchers use various analytical tools to obtain detailed quantitative data on subjects that have emerged during the course of fieldwork. For example, they may ask interviewees to rank fruits from most preferred to least preferred, rate the qualities of various types of firewood or sort a series of flashcards containing the names of animals into piles that reflect perceived similarities in morphology.

These exercises are often combined with interviewing techniques. Participants are asked to explain why they prefer a certain type of fruit over another or what the objects in each pile have in common. The tasks may be integrated into a survey, such as when local counterparts are asked to compare the properties of various medicinal plants in a structured interview about health care in the community.

Although each of the tasks described below is carried out in a slightly different way, there are some techniques and general advice that apply to all. Each analytical tool typically depends on the use of **props** that can be handled by the participants. These include dried or fresh specimens of plants, pictures of landscapes, notecards on which the local names of animals have been written and other visual aids which ensure that the interviewer and interviewee are referring to the same object. These

Table 4.5 Characteristics of various types of questions posed in anthropological research

	Appropriate stage of research	Information content	Ease of statistical analysis	Number of queries needed to explore subject	Breadth of subject covered
Open-ended	All stages	Very high	Low	Low	Very broad
Fill-in-the-blank	Middle to late	High	Medium	Medium	Broad
Multiple-choice	Late	Medium	High	High	Narrow
Dichotomous	Late	Low	High	Very high	Very narrow

props correspond to the members of a single **domain**, or class of objects, which is the focus of the research. A set of props might include, for example, a dozen types of firewood, ten kinds of bird skins or five varieties of corn.

Although tasks can be carried out without props, most researchers believe the best results are achieved with objects that respondents can see and touch rather than just think about abstractly.

Pay attention to factors that might affect the way people answer, including the limits on their attention span, the quality of the props and the order in which the items are presented. If the participants become bored, their answers will be less consistent than when they are focused on the task. Take this into consideration when choosing how many objects to include in each set, keeping in mind that for some tasks the amount of time needed increases rapidly in relation to the number of items included. The sequence of the objects should be randomized before each task to minimize the possibility that people's choices are influenced by the order in which they see the items. Cards can be shuffled, pairs of specimens presented in different sequences and the arrangement of the objects within each set changed.

Be careful that each prop you present is of good quality and is typical of the object you are seeking to represent. Let's say that you offer someone a choice between a green banana, a ripe apple and a rotting pear. Their selection will be likely to reflect their perception of the maturity of the fruits rather than preference of one species over the others. Blurry photographs, incomplete specimens and poorly written flashcards all detract from the quality of the data you will obtain. Pictures should show all the morphological characteristics needed to identify an object and differentiate it from other similar things. Biological specimens should contain most of the features that local people are accustomed to seeing. Combine specimens, pictures and other stimuli into a single prop if this increases people's ability to recognize the object. For example, show people photographs of birds, let them listen to recordings of the birds' songs and give them bird skins which they can feel with their hands. Mounted plant specimens can be accompanied by a drawing of the plant and a small bag of dried fruits or leaves that the participants can taste or smell.

Do your best to ensure there is no confusion over the identity of each object. If you are using name cards, be sure there is consensus about what is referred to by each label. As discussed elsewhere in this manual, many names are **polysemic** – referring to more than one object or concept. In addition, plant and animal names do not always have a one-to-one correspondence with scientific species, which means that the researcher may be thinking of one organism while the informant is visualizing another. If the task assumes that the respondents can recognize each specimen without difficulty, ask them to name and describe all the objects before ranking or sorting them.

If you are attempting to carry out the tasks in a scientifically rigorous way, interviews should be held with one individual at a time and with no onlookers present. This guarantees that the opinion of the interviewee is not swayed by the comments of others and that people who participate in the task in the future do not give pre-conceived answers.

There are both formal and informal ways of using the analytical tools presented below. In the informal approach, exemplified by participatory rural appraisal, it is customary to use small sets of items well known by all members of the community. These items can be sorted and characterized in a relatively short time. Researchers are casual in their approach to randomizing the order in which objects are presented and are relaxed about having people observe and comment on the tasks. At times, the tasks are administered in a single meeting to a large group of people who reach a consensus through discussion or who express their diverse opinions by voting. Props are used when available but often the exercises are carried out orally or written on a blackboard. Analysis of results is usually limited to summation of the responses, with little use of statistical approaches. The goal is to stimulate a dialogue with local people about their knowledge and management of resources or about other subjects which have been chosen by the community.

The formal approach, as pioneered by cognitive anthropologists, yields data which is analyzed using a variety of statistical measures. Psychologists, linguists and anthropologists who use this approach insist on randomizing the sequence of the objects and holding each interview in private. Emphasis is placed on the quality of the props. Although small subsets of items may be used at times, it is more common to carry out exercises which include all members of a certain domain or category. For tasks that focus on the classification or use of plant and animal resources, there can be hundreds of items that take many hours to sort or rank. The objective is to discover, in a scientifically rigorous way, how people perceive and classify the natural world.

4.3.6 Preference ranking

One of the simplest analytical tools involves asking people to think of some five to seven items in a category which is the focus of the research or of an issue which is being discussed in the community. Each person arranges the items according to personal preference, perceived importance in the community or another criterion. Each rank is given an integer value (1, 2, 3 and so on) with the most important or preferred item being assigned the highest number. For instance, in a set of five objects the most preferred is rated '5' while the least liked is '1'. These numbers are summed for all respondents, giving an overall ranking for the objects by the sample group of respondents.

Whenever possible, this order of preference is cross-checked with data obtained from interviews and other sources to see if there is consistency in the responses. In preference ranking of a few widely-recognized items, the task can be carried out orally or can be sketched on a large piece of paper which everyone can see. As the number of items grows it is preferable to have actual samples in hand and to randomize their order before asking each participant to rank them.

Preference ranking can be used, for instance, to elicit a list of plant resources that people feel are becoming increasingly rare in their communal forests. Each participant is asked to arrange the plants from least to most scarce. Another application of preference ranking is to ask people to make a list of the most valuable non-timber forest products in the region and then rank them in terms of the amount of income they yield.

Let's take the latter example, using the case study of Motisingloti village that is described in Box 6.4 (p.189). Shashi Kant and Neel Guarav Mehta reported that the total of some 126 000 rupees a year earned from selling forest products was derived from the following sources (100): timru leaves (39% of the total), mahua flower (23%), mahua seed (18%), bhindi seeds (15%) and sag seeds (5%). In Table 4.6, I have improvised answers for five fictional participants in a preference ranking exercise based on the income they earn from these forest products. The respondents rank the various plants in different ways but the overall order reflects the true income the village has earned from selling these resources.

4.3.7 Direct matrix ranking

Direct matrix ranking is a more complex version of preference ranking. Instead of arranging a series of objects on one characteristic such as 'value' or 'desirability', informants order them by considering several attributes one at a time. In other words, preference ranking is based on a single dimension whereas direct matrix ranking draws explicitly upon multiple dimensions.

After choosing a class of objects and defining its members, participants define the good and bad aspects of each. In the example given in Table 4.7, people might offer various observations – 'You can't make charcoal from palms, but you can from eucalyptus, acacias and pines', 'The fruits of eucalyptus aren't very useful' or 'Acacias yield good fodder'.

Table 4.6 An improvised preference ranking of the value of forest products for five households based on the case study of the economics of Motisingloti village in Gujarat, India (5, most valuable; 1, least valuable)

	Respondents					Total score	Ranking
	A	B	C	D	E		
Timru leaves	5	4	5	5	5	24	a
Mahua flower	4	5	3	4	3	19	b
Mahua seed	2	3	4	3	2	14	c
Bhindi seeds	3	1	2	2	4	12	d
Sag seeds	1	2	1	1	1	6	e

Table 4.7 The results of one person's direct matrix ranking of four tree species on seven use criteria, used as an example in participatory rural appraisal training in the Middle East and North Africa described by J. Theis and H.M. Grady (4, best; 1, worst)

	Eucalyptus	Palm	Acacia	Pine
Fuelwood	4	1	2	3
Building	4	1	2	3
Fruit	1	4	2	3
Medicine	4	1	3	2
Fodder	3	-	4	2
Shade	4	3	1	2
Charcoal	2	-	3	4
Total score	22	10	17	19
Rank	A	D	C	B

List these criteria along the left-hand size of a piece of paper and write the names of the items along the top. To simplify the scoring, rephrase negative statements as positive criteria. For example, 'The fruits of eucalyptus aren't very useful' is turned into a general criterion 'fruit' on which eucalyptus is given a low score.

Once the table is ready, people can rank the items according to each criterion, using a numerical scale in which the highest number is equal to the 'best' object and the lowest number to the 'worst'. The results of numerous individual responses can be added together to create a matrix that is representative of the community. Alternatively, direct matrix ranking can be done as a group exercise in which participants reach consensus on the ranking of each item or vote according to their individual assessments.

4.3.8 Triadic comparisons

In triadic comparisons, items are presented to participants in sets of three, as exemplified in Box 4.4. After defining the domain and identifying its most important members, triads are set up so that each item appears in every possible combination of three with the other items. After the sequence of the triads and the order within each triad are randomized, informants are shown each set of three and are asked to rank the items 'best to worst', 'most preferred to least preferred' or in another comparable way. Alternatively, the participants can judge the three items on overall similarity, selecting the item which is the most dissimilar to the others or the two that are most similar to each other. By posing questions such as 'Which fruit is the most dissimilar in taste to the others?', the interviewer can document patterns of similarity based on a particular attribute.

Because this test is more complicated than the previous ones, it is a good idea

Box 4.4 Judging the similarity of French pulses with triads test

In order to demonstrate the triads test for this manual, I purchased the following pulses (edible legume seeds) in a French market and assigned them numbers according to the alphabetical order of their names: (1) *chevriers* [the mature, dry seed of string beans (*Phaseolus vulgaris* L.)]; (2) *haricots rouges* [kidney beans (*Phaseolus vulgaris* L.)] (3) *lentilles blondes* [light-colored lentils (*Lens culinaris* L.)]; (4) *lingots blancs* [large navy beans (*Phaseolus vulgaris* L.)]; (5) *pois cassés* [split peas (*Pisum sativum* L.)]; and (6) *pois chiches* [chick peas (*Cicer arietinum* L.)]. I then made a table of all possible combinations of three for these pulses (Table 4.8).

Table 4.8 Alphabetized list of all possible combinations of six types of legumes into sets of three

Number	Order	Items
1	1,2,3	Chevriers, Haricots rouges, Lentilles blondes
2	1,2,4	Chevriers, Haricots rouges, Lingots blancs
3	1,2,5	Chevriers, Haricots rouges, Pois cassés
4	1,2,6	Chevriers, Haricots rouges, Pois chiches
5	1,3,4	Chevriers, Lentilles blondes, Lingots blancs
6	1,3,5	Chevriers, Lentilles blondes, Pois cassés
7	1,3,6	Chevriers, Lentilles blondes, Pois chiches
8	1,4,5	Chevriers, Lingots blancs, Pois cassés
9	1,4,6	Chevriers, Lingots blancs, Pois chiches
10	1,5,6	Chevriers, Pois cassés, Pois chiches
11	2,3,4	Haricots rouges, Lentilles blondes, Lingots blancs
12	2,3,5	Haricots rouges, Lentilles blondes, Pois cassés
13	2,3,6	Haricots rouges, Lentilles blondes, Pois chiches
14	2,4,5	Haricots rouges, Lingots blancs, Pois cassés
15	2,4,6	Haricots rouges, Lingots blancs, Pois chiches
16	2,5,6	Haricots rouges, Pois cassés, Pois chiches
17	3,4,5	Lentilles blondes, Lingots blancs, Pois cassés
18	3,4,6	Lentilles blondes, Lingots blancs, Pois chiches
19	3,5,6	Lentilles blondes, Pois cassés, Pois chiches
20	4,5,6	Lingots blancs, Pois cassés, Pois chiches

Since the triads in this table have an artificial order that might affect the results of the study, I randomized the sets in two ways: (1) by changing the sequence in which the triads are presented and (2) altering the order of the items within each triad. For the first step, I referred to the random numbers table presented in Figure 4.3. Starting at the beginning of the third row and moving downward, I scanned the first two digits for numbers between 1 and 20, which correspond to the numbers of the triad sets in Table 4.8. Because not all the combinations were found after consulting the first two numbers,

I continued the search using the middle two numbers and then the final two numbers, always beginning at the top of the third column. I obtained the following arbitrary order: 13; 01; 08; 02; 06; 03; 09; 14; 16; 19; 05; 20; 17; 12; 18; 11; 07; 10; 15; 04.

For the second step, I used slips of paper numbered 1–6 to randomize the order of the objects in each set. I decided arbitrarily that if the slip of paper came up (1), the order would be 1st object, 2nd object, 3rd object, based on the alphabetized table; and that (2) = 2nd object, 1st object, 3rd object; (3) = 3rd object, 1st object, 2nd object; (4) = 1st object, 3rd object, 2nd object; (5) = 2nd object, 3rd object, 1st object; (6) = 3rd object, 2nd object, 1st object. When I pulled slips of paper for this task, I had the following results: 4; 1; 4; 6; 2; 2; 5; 5; 4; 5; 3; 6; 1; 3; 3; 5; 6; 6; 6; 1.

Table 4.9 A fully randomized list of all possible triads of six types of pulses. One informant's assessment of the most dissimilar object of each set is indicated in **bold**

Number	Order	Items
13	2,6,3	*Haricots rouges, Lentilles blondes,* **Pois chiches**
01	1,2,3	*Chevriers, Haricots rouges,* **Lentilles blondes**
08	1,5,4	*Chevriers,* **Pois cassés,** *Lingots blancs*
02	4,2,1	*Lingots blancs,* **Haricots rouges,** *Chevriers*
06	3,1,5	*Lentilles blondes,* **Chevriers,** *Pois cassés*
03	2,1,5	*Haricots rouges, Chevriers,* **Pois cassés**
09	4,6,1	*Lingots blancs,* **Pois chiches,** *Chevriers*
14	4,5,2	*Lingot blancs,* **Pois cassés,** *Haricots rouges*
16	2,6,5	*Haricots rouges,* **Pois chiches,** *Pois cassés*
19	5,6,3	*Pois cassés,* **Pois chiches,** *Lentilles blondes*
05	4,1,3	*Lingots blancs, Chevriers,* **Lentilles blondes**
20	6,5,4	**Pois chiches,** *Pois cassés, Lingots blancs*
17	3,4,5	*Lentilles blondes,* **Lingots blancs,** *Pois cassés*
12	5,2,3	*Pois cassés,* **Haricots rouges,** *Lentilles blondes*
18	6,3,4	**Pois chiches,** *Lentilles blondes, Lingots blancs*
11	3,4,2	**Lentilles blondes,** *Lingots blancs, Haricots rouges*
07	6,3,1	**Pois chiches,** *Lentilles blondes, Chevriers*
10	6,5,1	*Pois chiches,* **Chevriers,** *Pois cassés*
15	6,4,2	**Pois chiches,** *Lingots blancs, Haricots rouges*
04	1,2,6	*Chevriers, Haricots rouges,* **Pois chiches**

Table 4.9 shows the resulting randomizations for the sequence of triads and for the order within each triad. Once the fully randomized table was ready, I showed each triad of objects to the editor of this manual and asked him to indicate the most dissimilar item. His responses are shown in bold in Table 4.9. Because each pair of legumes appears a total of four times in the full set of 20 triads, they can be ranked in terms of similarity on a scale of

0–4. These results, presented in the matrix shown in Figure 4.5, reflect the morphological similarity (or dissimilarity) of the pulses. The three species of *Phaseolus vulgaris* are frequently matched because they all have kidney-shaped beans. The split peas and the lentils are grouped together as much for their similar size and rounded shape as for their dissimilarity to beans and chickpeas. The chickpeas are in a class of their own. This result is confirmed by observing that chick peas were considered the most dissimilar item nine out of the ten times they appeared in the triads. The other legumes were considered to be the most dissimilar item between one and three times.

The results of triadic comparisons by different individuals can be added together to obtain an aggregate matrix. The combined responses, which give a quick idea of the majority opinion in the community, can be further analyzed with sophisticated statistical techniques such as multidimensional scaling and clustering to reveal patterns of agreement.

Keep in mind that triadic comparisons can also be used as a tool for ranking. Every item that appears in the set of three is given a rating of 1, 2 or 3, with three being the highest. The ratings of each item are added together to obtain the overall rank. In the case of the six French pulses, each item appears ten times in the set of 20 triads and would thus have an overall score which ranges between 0 and 30.

Triads are usually carried out with relatively few items because the time needed to carry out the task increases exponentially as additional objects are added. The total number of triads needed is given by the formula $n!/3!(n\text{-}3!)$, where n is equal to the number of items. A number followed by ! is referred to as a **factorial** and is obtained by multiplying the series of integers between the number and 1. Thus, 3! is computed as $3 \times 2 \times 1 = 6$, whereas 6! is calculated $6 \times 5 \times 4 \times 3 \times 2 \times 1 = 720$. The example of six pulses requires 20 triads $[6!/3!(6–3!) = 720/36 = 20]$, whereas eight items would necessitate 56 triads $[8!/3!(8–3!) = 40\,320/720 = 56]$ and ten items would result in 120 triads $[10!/3!(10–3!) = 3\,628\,800/30\,240 = 120]$. This rapid escalation in the number of triads means that presenting more than ten items in a single triadic comparison is impractical. A variation on this technique, the balanced incomplete block design, allows up to 25 items to be compared in triads in a reasonable amount of time.

to begin each exercise with a few sample questions so that the local participants can become familiar with the procedure. These questions should be drawn from a subject different from the one you are exploring so that the answers do not affect the participants' performance on the real triadic comparison. For example, if you are testing for similarity of firewoods or preference for different fruits, begin the task with some triads of ornamental flowers, seed types or some other familiar object. Once people understand how the triadic comparison works, start the exercise using the items that you have selected.

Some interviewers ask respondents to explain, either during the task or after it is finished, the choices they make in specific triads. Depending on whether the triads are being used for ranking or for assessing similarity you can pose questions such as 'Why do you think these two items are similar?', 'What makes this item different from the others?' or 'Why do you prefer this item over the other two?'

The responses for similarity comparisons can be tallied in a matrix such as the one in Figure 4.5. For preference ranking, an overall ranking for each item is obtained by adding together the ranks it is given in each triad in which it appears. In a triadic comparison of four items such as apples, oranges, peaches and pears, each object appears in three different triads. If the pears were rated as '3', '2' and '3' in the three triads in which it appears, its overall rating would be '8'.

4.3.9 Paired comparisons

In paired comparisons, interviewees are shown items which have been arranged in sets of two. Based on discussions with the participants or on previous interviews in the community, define five to ten important options or objects which characterize the subject to be explored. Make written notecards for each option or, in the case of objects, prepare good quality specimens.

	Chevriers	Haricots	Lentilles	Lingots	Pois cassés
Chevriers					
Haricots	3				
Lentilles	1	1			
Lingots	4	3	1		
Pois cassés	0	1	4	1	
Pois chiches	0	0	0	0	1

Figure 4.5 Tabulation matrix for triads test using six pulses. The number of matches between any two items ranges from 0 to 4.

Box 4.5 Paired comparisons of some common fruits

In order to provide an example of paired comparisons, I asked the editor of this manual to state his preference for five common fruits – apple, banana, grapefruit, orange and pear – which I showed him in pairs. Starting from an alphabetical list, I randomized both the sequence of the pairs and the order within each pair. The sequence can be decided by numbering the pairs from 1 to 10 when they are in alphabetical order and then reordering them by reference to a random numbers table or by pulling numbered slips out of a hat, as previously explained. The order within each pair is given by flipping a coin. If heads, the original order is maintained but if tails, the items are switched around. The results of this reordering and the editor's choices are shown in Table 4.10. The responses can be tallied by hand or in a pairwise matrix such as the one shown in Figure 4.6.

Table 4.10 Paired comparisons for five fruits that have been randomized for the sequence of the pairs and the order within each pair. The informant's preference within each pair is shown in **bold**

Pair	Order	Items
9	5,3	**Pear**, Grapefruit
5	3,2	Grapefruit, **Banana**
6	2,4	Banana, **Orange**
8	4,3	**Orange**, Grapefruit
10	5,4	Pear, **Orange**
1	1,2	**Apple**, Banana
4	1,5	**Apple**, Pear
2	1,3	**Apple**, Grapefruit
7	5,2	Pear, **Banana**
3	4,1	Orange, **Apple**

Apple	Banana	Grapefruit	Orange	Pear		Score	Rank
	Ap	Ap	Ap	Ap	Apple	4	A
		Ba	Or	Ba	Banana	2	C
			Or	Pe	Grapefruit	0	E
				Or	Orange	3	B
					Pear	1	D

Figure 4.6 A pairwise ranking matrix for five fruits.

By creating a chain of preferences based on the ranking, you can see if the relationships of preference are consistent throughout the exercise. Given the results in Table 4.10, we can arrange the fruits as follows: Apple > Orange > Banana > Pear > Grapefruit. There is perfect **transitivity** in these results, which means there are no logical contradictions such as finding that the informant likes bananas better than pears and pears better than grapefruit, but grapefruit better than bananas. Perfect transitivity adds credibility to the data. Inconsistent results would lead you to believe the answers are arbitrary, people do not have strong preferences, the objects are classified on multiple conflicting dimensions or that pairwise ranking is not an appropriate analytical tool for the cultural domain tested.

Adding responses of a representative group of informants allows you to make statements on general cultural trends. You can do this quickly by adding the overall number of times a fruit is chosen by each informant, as reflected in the 'score' column of the pairwise ranking matrices. A more elaborate method is to add together the results within each cell of the matrix, thus breaking down the overall score into its components. These combined results will allow you to state community preferences in a way that is statistically significant. The aggregate scores can be analyzed with a statistical tool called **clustering** to discover if patterns of preference exist within the community. Clustering techniques allow you to evaluate the similarities and differences in the way that informants respond.

Pairwise ranking is normally used with relatively few items because the time needed to carry out the task increases exponentially as you add additional objects. The total number of pairs required is given by the formula $n(n-1)/2$, where n is equal to the number of items. The example of five fruits required 10 pairs $[5(5-1)/2 = 10]$, whereas 10 items would necessitate 45 pairs $[10(10-1)/2 = 45]$ and 15 items would yield 105 pairs $[15(15-1)/2 = 105]$. The total number of items you choose will depend not only on the complexity of what you are studying but also on the patience of the participants.

Before you begin, compose either a list or a matrix of all possible combinations, as discussed in Box 4.5. This will allow you to keep track of the responses and to ensure that all pairs are shown during the course of the task. After randomizing the order of the pairs, present them to the interviewee, who will choose the one he or she prefers or considers most important – 'the best fruit', 'the highest quality firewood', 'the major cause of deforestation' or another response that is appropriate for the question posed.

In order to gain insight into people's reasoning, you can ask the participants to describe why one option is better or worse than another. In addition, you can ask if the preferred item has any negative qualities or if the one not chosen has any positive aspects. Some researchers ask for these comments after each choice, whereas others prefer that respondents complete the entire task before giving their general observations on the overall pattern that emerges. If you ask someone to say why they like one fruit better than another after each choice, they might respond with statements such as 'grapefruits are too sour' or 'bananas are too slimy but I like their flavor' or 'pears taste good because they are sweet and juicy'. If you asked for their opinion after finishing the exercise, they may be able to synthesize these observations into a general statement along the lines of 'I like sweet, juicy fruits but I can't stand sour or slimy ones'. Although these conclusions may be obvious for some domains – including favorite fruits – they often allow both the interviewer and the respondent to discover underlying rules or tendencies that explain people's behavior.

After the task is done, you can ask the participants if they have thought of any additional items or criteria left out of the original listing. If so, these can be merely inserted at the appropriate place in the ranking or you can create a new set of possible pairs for the informant to review and comment upon.

On another day, you can verify the consistency of the answers by asking respondents to carry out a preference ranking with the same fruits. A simpler way to check is to pose directly the question, whether in an interview or in open conversation, that was addressed in the exercise: 'What is your favorite fruit?' or 'What is the greatest cause of deforestation in the community?'.

4.3.10 Pile sorting

In the pile sort task, participants divide objects into a number of groups according to the overall similarity of the items. Imagine that someone asks you to sort the produce from a home garden into piles containing similar objects. You might start by creating separate mounds of fruits, vegetables, medicinal herbs and ornamental flowers. If you were asked to continue, you might divide the fruits into a series of piles that contain berries (raspberries, strawberries, blueberries), citrus fruits (oranges, grapefruits, lemons) and pome fruits (apples and pears). Finally, you may make distinct piles for oranges, grapefruits and lemons. This would be continued for the other fruit, vegetable, medicinal herb and flower piles until all objects are classified.

In practice, pile sorts are usually carried out with cards, which allows large numbers of concrete or abstract things to be classified in a relatively short period of time. After defining the domain that interests you, write the name of all relevant items on small notecards, using large and legible print. If you are working with non-literate informants, you will be obliged to use picture cards or real objects, ensuring that they include the key features that people use to recognize and classify the items.

Each participant shuffles the cards and then proceeds to make piles according to his or her notion of similarity. In some cases you may want to specify that the items be sorted according to a certain attribute that interests you, such as commercial value, scarcity or utility. Some researchers allow as many piles as the interviewee chooses, whereas others workers restrict the number by asking 'Please sort

Carrots	Corn	Oranges	Bananas
Potatoes		Lemons	
Beets			

| ——— | ——— | ——— | ——— |
| Pile 1 | Pile 2 | Pile 3 | Pile 4 |

Figure 4.7 A pile sort of produce harvested from a home garden.

the cards into two (or three, four or more) piles of similar objects'. You can obtain additional information that will help you to interpret the results by asking the participants what the objects in a single pile have in common or what differences there are between two separate piles. Most people do this after all the piles have been formed so that the participants do not change the way they sort in the middle of the exercise.

After the piles have been made, the results are tabulated in a similarity matrix like the ones used in triadic and pairwise comparisons. In order to demonstrate this, let's return to the example of home garden produce with which this section began. If you were asked to sort out a harvest of bananas, beets, carrots, corn, lemons, oranges and potatoes into an unlimited number of piles, you might come up with the arrangement shown in Figure 4.7.

Draw a matrix, such as the one depicted in Figure 4.8, that gives a pairwise comparison of all these items. Pairs found in the same pile are scored with a one. If the two items are found in different piles, the corresponding box in the matrix is filled with a zero. As additional people complete this task, community-wide patterns of the perceived similarity between the objects will emerge which can later be analyzed using various statistical approaches. By comparing the matrices presented in Figures 4.5, 4.6 and 4.8, you can see that triadic and pairwise comparisons yield more data per person than sorting tasks. For this reason, 20 or more participants are needed in pile sorting to achieve statistically reliable results, whereas less than 10 are acceptable in triadic and pairwise comparison tasks.

There are many variations on the pile sort that add flexibility to the technique, allowing you to adapt it to your own study. For example, if someone believes that

	Bananas	Beets	Carrots	Corn	Lemons	Oranges
Beets	0					
Carrots	0	1				
Corn	0	0	0			
Lemons	0	0	0	0		
Oranges	0	0	0	0	1	
Potatoes	0	1	1	0	0	0

Figure 4.8 A similarity matrix based on the results of a pile sort of fruits and vegetables harvested from a home garden.

133

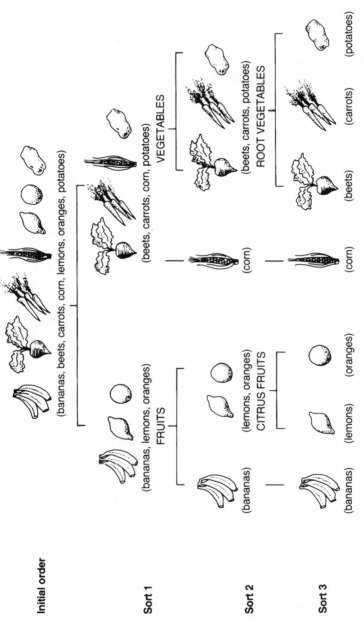

Figure 4.9 Taxonomic structure derived from a pile sorting of home garden produce. Piles are indicated by parentheses ( ) and named subgroupings are shown in capital letters.

an item belongs in more than one pile, additional cards can be made and placed in the various groupings, effectively allowing people to split their vote. During the course of the exercise, people may think of additional items that were left out of the original set of cards. These can be written onto cards and included in the appropriate piles at any point during the task. After participants make a first sort, you can ask them to break down the resulting piles into ever smaller ones. This successive sort is scored in the same way as a single sort. Pairs that remain together in the same pile after the second sort are given a value of '1', which is added to the value in the appropriate matrix box. If they are still together after the third stage of sorting a '1' is again added and so on until they are split into different piles or the sorting task ends.

If the sorting continues until every pile contains just a single item, it gives a complete taxonomy of the items which can be sketched out on paper. As you construct this tree with local counterparts you can ask if the various subgroupings have names. An improvised example of a complete successive sort and a resulting taxonomic structure is given in Figure 4.9. This technique is particularly useful in studies of local classification of plants and animals, a subject that is discussed in greater detail in Chapter 7.

5

Ecology

Figure 5.1 Edward Mabogo of Venda, South Africa, measuring the diameter of a tree during an ethnoecology course in Tepoztlán, Mexico. The tape is placed at breast height, avoiding irregularities in the trunk.

5.1 Describing microenvironments and quantifying their plant resources

As with other approaches to ethnobotany, ecological studies have both qualitative and quantitative elements. Researchers often begin by describing the different ecological zones, or **microenvironments**, recognized and in many cases transformed by local people. These zones are characterized by distinct soils, vegetation and other components which are themselves further classified by local people and scientists. After classifying the microenvironments, ethnobotanists often make a quantitative assessment of the plant resources in each zone.

5.2 Qualitative approaches

Local botanical knowledge is interrelated to perception of landforms, soils, climates, vegetation types, stages of ecological succession, land use and many other aspects of the natural environment. Each of these subjects has been studied in depth by scientists and there is a growing literature on how they are classified by local people.

With the exception of soils, samples of these domains are not collectable – you cannot take away a landform or put ecological succession in a museum. Documentation depends on excursions into the field with knowledgeable informants, who can help you visualize the environment through their own eyes. Along a particular slope, a local person may be able to indicate the different stages of ecological succession or point to vegetation types which are visible from afar. Photographs can be taken of these various dimensions and shown to other people in the community for verification. Another approach is to draw a map of the landscape, later filling in climate zones and vegetation types. These field observations and visual stimuli can be used to compare notes with soil scientists, climatologists and ecologists who work in the region. In the following sections, I compare the local and scientific ways of looking at the environment and describe the methods used to elucidate the correspondence between the two. The principal dimensions to consider when studying each domain are summarized in Table 5.1.

5.2.1 Landforms and geographical localities

A good way to start an ethnobotanical study is to understand how local people classify rivers, hills, valleys and other geographical features which dominate the landscape. You will find that there are general names for each **landform**, a feature of the earth's surface which has been formed by natural forces. Communal lands are usually delimited by rivers, mountains and ridges which have political and symbolic importance. Over the years, these localities have been given specific place-names, or **toponyms**, by the residents.

In order to grasp how landforms are classified, go to an open area with some people from the local community. For example, imagine that you accompany a

Table 5.1 A summary of various domains of folk ecological knowledge and some of the principal dimensions on which they are classified

Domain	Dimensions
Landforms	Elevation, location and shape; includes hilltops, rivers, valley bottoms, plateaus, cliffs and many other forms
Soils	Color, texture, fertility, acidity–alkalinity, workability, humidity, consistency, drainage, profile
Climate	Temperature, rainfall, evapotranspiration, elevation, exposure, topography (including position of land masses and bodies of water), winds, seasonality
Vegetation types	Physiognomy, floristic composition (including dominant species), parameters of the habitat, extent of human management or disturbance
Stages of ecological succession	Floristic composition, height of vegetation, regeneration under different environmental conditions, life histories of individual species, number of years left fallow, current human management, extent of original disturbance
Land use zones	Produce (including tangible goods and intangible benefits), technology applied (including amount of human labor invested), extent of management of the environment, distance from household or community, ownership; all dimensions mentioned under other domains above are potentially included

group of Sahaptin-speakers to a river plain in the Pacific Northwest of the United States, overlooking the landscape depicted in Figure 5.2.

Look around and let your questions flow naturally from what you see: Does the river there have a name? How about that hilltop? Is there a general word for river, mountain, valley bottom and other geographical features that dominate the landscape? When you discover a place-name, ask about its meaning and origin. Many names can be attributed to local plants or animals that are in abundance, to the past or present (or mythical) owners of the land or to a salient feature of the landform. Some toponyms are related to important events described in legends or to something that happened in recent history. Yet other names will be obscure, their meaning lost with the passage of time.

Another method complements this first brief glimpse of the topography. Ask the villagers to draw a map of their community, indicating all geographical boundaries and important landforms. After finishing this general map, you can work with them on more detailed diagrams of specific features such as river systems. You may also choose to focus on a smaller area of the community. There will be a high density of named localities in the community itself and in nearby cultivated areas but fewer and fewer in wildlands. This is because the number of place names

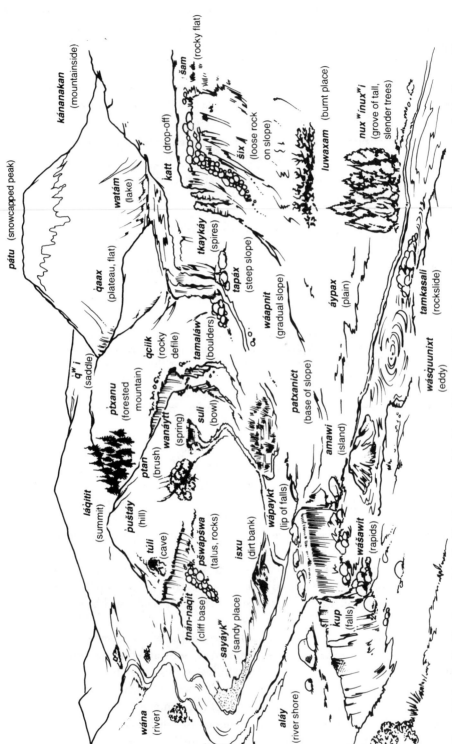

Figure 5.2 An idealized landscape of the Pacific Northwest of the United States, with Sahaptin names for various landforms as recorded by Eugene Hunn.

is related to the intensity with which the land is used and to its value to local people.

Whenever possible, make field visits to these localities accompanied by people from the village. Bring along an altimeter to measure the altitude and estimate the latitude and longitude with topographical maps or a geographical positioning system. Make notes on the vegetation, soil type, climate and other characteristics of the site as perceived by the villagers. You can later compare these data, your notes and the village-made maps with topographic maps and satellite imagery. This will help you orient yourself and locate the landforms that have been mentioned. As a further step, you can show the technical maps and images to villagers to see if they can furnish any additional information on the topography. Do not assume that the abstract bird's eye view projected in such figures will correspond directly to the local way of classifying landforms. As Eugene Hunn noted for the geography of the Columbia Plateau depicted above [74]:

> For non-Indians, a focus on specific mountains and rivers as things of importance implies a cartographic perspective, one in which the observer is placed above the landscape as if in flight. The Indians' land-based perspective named instead specific places on a mountain or along a river *where things happened*. It was a practical rather than a purely abstract geography, naming culturally significant places, the sites of important events or activities, whether of the present or of the myth age.

5.2.2 Soils

Soil scientists study the origin, composition and quality of soils, which they characterize on a variety of dimensions [75, 76]. They pay attention to physical properties such as **texture** (determined by the size and number of component particles) and **structure**, the way that these particles are arranged in the soil. Soil researchers also measure the diverse organic and chemical constituents in the soil. After taking samples from several different locations, they carry out laboratory analyses to discover the relative contents of organic matter, which is derived from decomposed plants, inorganic matter, and primarily consisting of mineral particles such as sand, clay, silt and gravel. In addition, they measure the contents of nitrogen, phosphorus, potassium and trace minerals that are required for plant growth. Finally, they test the pH of the soil, revealing if it is acid, alkaline or neutral. Together, the composition and physical properties give the soil a distinctive color and consistency which are taken into account in most systems of soil classification.

One of the major advances in soil classification was the recognition of **horizons**, the layers observed by making a vertical cut that exposes the soil from the earth's surface down to a depth of 1–3 m, sometimes reaching bedrock. This vertical slice of the earth is called the soil **profile**.

Color, texture, consistency and utility appear to underlie indigenous categories in many parts of the world. Secondary characteristics, such as humidity, drainage, salinity, fertility and workability are important in some systems. Local people may also observe the different layers that make up the soil profile, giving special attention to the relative depth of the topsoil, the presence or absence of a claypan layer which inhibits the growth of roots and the type of parent material from which the soil developed. They may also note the amount of living matter in the earth, including plant roots, worms, insects and burrowing animals. The relationship between distribution of soils and topography is often well understood, including the relative fertility of the earth near rivers, at the base of slopes and in certain forested areas. The susceptibilities of different types of soil to erosion and leaching are often referred to in indigenous classifications. Often, the quality of the soil is judged by observing the covering vegetation. Indicator species reveal the location of nutrient rich soils or sites where the earth is leached or eroded.

After you have elicited local soil categories and understood their general characterisitics, go to the field with someone who can locate areas where each of these soils is found. Record the name of each locality and take notes on the covering vegetation, human use and its location relative to nearby landforms (such as whether it is at the top of a hill or in a valley bottom). Ask how each soil type is distinguished from others, keeping in mind the primary and secondary characteristics mentioned above. Some soil features can be further documented using simple tools. Soil color charts allow you consistently to record the various shades of browns, reds, ochres and other hues that are encountered. Additional equipment is available for gauging moisture, texture and chemical properties such as pH. In more advanced studies, an auger is used to extract soil cores which reveal the depth and characteristics of each horizon.

The most useful equipment of all in soil studies are shovels and bags, which allow you to prepare samples and transport them to soil scientists working at well-equipped laboratories. After clearing accumulated plant litter, remove a shovelful of soil from the top horizon. Although you should dig deeply enough to get a representative sample, avoid mixing soil from different horizons – the top soil may only be a few centimeters deep in some areas. This is then mixed or **homogenized**, to ensure that the soil particles and humidity are well distributed. Several handfuls of the soil are placed in a durable plastic bag, which after sealing is ready for the laboratory. This is repeated for the other horizons of the soil profile. Alternatively, collect samples at standard intervals, say 0–5 cm and 25–30 cm, beneath the soil surface or litter; consult soil scientists for details of sampling methods appropriate to particular studies.

The soil scientists who examine these samples will produce a report which includes the various characteristics discussed above. If the soil comes from a productive area such as an agricultural field, they will also suggest which nutrients are deficient, limiting the growth of crops or managed plants. Finally, they will classify the soil according to the system most commonly used in local surveys. With this report in hand, a rigorous comparison can be made between the scientific and indigenous ways of looking at soil and judging its productivity.

5.2.3 Climates

Climatologists set up weather stations in order to measure rainfall, temperature, evaporation and other aspects of the weather. Because they take records every day for many years, they are able to observe not only seasonal fluctuations but also variations from year to year. This information is generally available in the form of weather tables organized by locality. For many regions, maps have been made which show bands of average rainfall and temperature, usually correlated to elevation and other topographical features.

Agriculturalists and other local people, dependent upon favorable temperature and rainfall for a successful harvest, are careful observers of the climate. They often have elaborate methods of predicting changes in the weather from one year to the next, based on observations of the atmosphere, the behavior of animals, the timing of temperature changes and the onset of rainfall. These phenomena are closely related to the agricultural calendar, which governs when to prepare the soil, sow seed, harvest and perform other tasks. Local people also have a keen sense of microclimates, the small localized variations in climate which allow a particular vegetation or cultivated plant to prosper.

Although climate is not something that can be seen or touched, everyone is aware of it. Yet if you cannot point to a climate zone and ask its name, how do you elicit the local categories? You will probably hear some climate terms mentioned in everyday conversation, especially when people are talking about agricultural production or the location of a settlement: 'This corn grows in cold country' or 'His ranch is in the hot zone'. Because each term is likely to have a common word such as 'country', 'place' or 'land', you can ask for other similar expressions. For example, I learned that the village of Totontepec is located in *mukojk an it*, the 'temperate zone'. After enquiring for other zones, I was told that 'hot land' was called *an it* and 'cold country' *xox it*.

After obtaining a complete list of names, ask local people to draw a map showing the limits of each climate zone in the community. Local people characterize broad climatic zones by temperature ranges and rainfall. They can make a diagram of how these and other factors vary seasonally in each climate zone. With community weather records or climate and topographic maps in hand, you can estimate the temperature, precipitation and elevational limits of each area.

If you lack access to maps and records, you can use an approach called seasonality diagramming, which is borrowed from participatory rural appraisal. Define with local people when their year begins and how they divide time (into seasons or months, for example). Next, verify which dimensions – rainfall, temperature, windiness or sunshine – are used to characterize the climate throughout

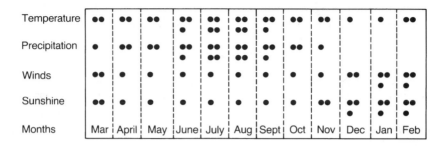

	Mar	April	May	June	July	Aug	Sept	Oct	Nov	Dec	Jan	Feb
Temperature	••	••	••	•••	••••	••••	•••	••	••	•	•	••
Precipitation	•	••	••	•••	••••	••••	•••	••	•			
Winds	••	•	•	•	•	•	•	•	•	•••	•••	•••
Sunshine	••	•	•	•	•	•	•	•	••	•••	•••	•••
Months	Mar	April	May	June	July	Aug	Sept	Oct	Nov	Dec	Jan	Feb

Figure 5.3 An improvised diagram of the seasonal variation in various dimensions of climate.

the year. Mark these dimensions and time units on a piece of paper or on the ground to create a diagram such as the one in Figure 5.3. Villagers can indicate the relative quantity or degree for each characteristic. For example, they may put one dot or stone in the temperature box for cold, two for warm and three for hot. No dots in the precipitation category would indicate a dry period and four dots would mark the wettest period. A diagram should be made for each climate zone recognized by the villagers. The separate drawings can later be unified into a single one, which is correlated with topography and elevation.

5.2.4 Vegetation types

Classification of vegetation has stirred much controversy. The disagreement begins with whether or not distinct units of vegetation exist. Some ecologists think that vegetation types represent a continuum, that it is an unhelpful abstraction to recognize named units and that grouping them into associated segments is largely a figment of scientific imagination. Despite this scepticism, many researchers agree that such units can be recognized in nature. As Robert McIntosh, an historian of ecology, noted [77]:

> The long tradition of natural history, the ordinary experience of farmers, seamen, woodsmen, hunters, anglers and herbalists, indeed, of the earliest of food gatherers and the plethora of words in many languages describing particular kinds of aggregates of organisms and often their associated habitats, testifies that the earth is covered with a complex pattern of 'more or less' recognizable plant and animal communities.

Natural vegetation types used to be strictly differentiated from artificial ones. Ecologists dedicated themselves to the study of natural vegetation – including forests, mangroves and grasslands – whereas agronomists focused on intensively cultivated sites such as agricultural fields, cattle pastures and plantations. Practitioners of the emerging fields of agroecology and human ecology seek to unite these two perspectives. Today, there is growing recognition that, generally, even the apparently most pristine forests have been greatly affected by humans, and that many cultivated areas are integrated into the surrounding vegetation and include non-domesticated species.

Characterizing natural vegetation has been a difficult task, given the large number of factors which influence the formation of plant and animal communities. When describing a vegetation type, many researchers begin by defining the **habitat**, or home territory, of a group of associated living organisms. The habitat is characterized by specifying the variability of numerous ecological and geographic parameters including latitude, soil, elevational range, topography, average temperature and rainfall as well as the interaction between plants and animals.

Many classification systems are based on the **physiognomy** or general appearance of the vegetation. A broad range of structural and functional attributes has been used to define appearance, including the height and number of **strata** – the vertical layers formed by forest plants of similar height – as well as the overall color and luxuriance of the foliage and the size and shape of typical leaves. Note is also made of the relative proportion of different lifeforms and any periodic cycles of leaf emergence and leaf fall, flowering and fruiting.

Other classifications are derived from **floristic composition**, the assemblage of plant species that characterize a vegetation type. The simplest analyses simply give a qualitative assessment of the most common plants, called **dominant species**. More detailed studies are based on counting the plants within a predetermined sample area. The most common measures of floristic composition are **richness**, the number of different species, and **abundance**, the number of individuals per species found in a specified area. In tropical forests, these studies have demonstrated that there is usually a high diversity of tree species, but a low density of most species. In temperate areas, there is a lower diversity of forest trees and some species are very abundant.

Parallel to classifications of natural vegetation, agronomists recognize diverse cultivation systems. Their primary criterion is the dominant crop that is grown at a site, but they also differentiate between areas planted as a monoculture (a field containing a single species) or as a polyculture (a field containing numerous crops). In addition, the habit of the cultivated plants is taken into account. The majority of zones have herbaceous plants which may be either annual (replanted every year) or perennial (a single planting lasting for many years). Other areas are covered by woody plants (orchards or coffee groves) or by vines (grape vineyards). Special notice is taken of the water source of the cropland – be it rainfall or from one of a number of different systems of irrigation – and if the plot is permanently farmed or left to fallow for a number of years in between cycles of cultivation.

Agroecologists and human ecologists seek to understand how wild and cultivated zones of vegetation coexist. They focus on polycultures, in which weedy and semi-cultivated species are permitted to grow among domesticated plants, and on agroforestry systems, in which useful trees and shrubs are intermixed with culti-vated herbaceous plants.

Much like agroecologists, local people consider that cultivated areas, aban-doned fields, shrublands and forests form a continuum of vegetation types. Vege-tational units, distinguished by physiognomy, habitat and composition of the flora, are often labeled with the name of the dominant or most culturally significant plant at the site. Among the Mixe, both wild and cultivated vegetation are characterized as types of **kam**, which may be loosely translated as *field* or more literally as *place of*. Each kind of **kam** is identified by the dominant species cover. For example, grasslands are called **tsoots kam** *grass place*, oak forest is **xojkam** *oak place*, a blackberry patch is **tsa'am ju'u kam** *blackberry place* and a fern meadow is **tsimkam** *fern place*. Cultivated fields follow this same pattern of naming. The *milpa* or cornfield, is called **mokkam** *corn field*, a sugarcane plantation is referred to as **vaxkam** *sugarcane field*, a coffee grove is **cafekam** *coffee field* and so on. These vegetation types and agroecosystems are characterized not only by the dominant species but also the many cultivated and non-cultivated plants that grow in association.

Eliciting the classification of vegetation types can proceed in much the same way as a study of landforms. From afar, local people can indicate the names of the different associations and sketch their size and location on a hand-drawn map. As in the Mixe example given above, you may find that these units are designated by a binomial composed of a specific term referring to the dominant species and a general term for place or locality. Once you understand the logic of the naming system, you can try out other names to see if they are commonly used by local people: Is there an oak-place?, a potato-place? a blackberry field? Some terms will sound ridiculous to people, others plausible and others familiar. This will allow you to estimate the number of vegetation terms that exist and their relative importance to local people.

After compiling a list and seeing the vegetation types from afar, you can venture into them to record your observations on the species composition and structure. Eko Walujo, an Indonesian ethnobotanist, used this method to characterize various vegetation types recognized by the Dawan people of Timor [78]. Working in plots of two types of forest vegetation – **nasi** *natural stands* and **kiuk kotok** *sacred groves* – he counted the number of individual woody plants and determined each to species. In the **nasi** sites, he found 18 species of large trees and 23 species of shrubs and small trees. His survey of three sites in the **kiuk kotok** yielded 27 species. Species lists from the two forest types showed only partial overlap and the trees common to both sites were found in different densities. Additional surveys revealed even greater differences in species composition and physiognomy be-tween the human-made vegetation types such as 'private and communal agricul-tural fields' (**lele** and **po'an**), 'private and communal gardens' (**kintal** and **kuan**) and natural areas such as 'savannahs' (**hu sona**).

5.2.5 Classification of ecological succession

Plant ecologists have long been interested in the dynamics of vegetation growth. They observe the **succession** of species that appear and disappear in a particular place over time and note the resulting changes in the physiognomy of the vegetation. They gauge the number of years that it takes for mature forest to develop on a site that has been freshly colonized (**primary succession**) or to return to an area that has been disturbed by natural causes or humans (**secondary succession**).

In recent years, much research has focused on **regeneration** – regrowth of natural vegetation – along the forest edge and in the gaps created by tree fall. Direct sunlight penetrates these areas, causing changes in temperature, humidity and soil moisture which stimulate seeds to germinate and dormant seedlings to begin accelerated growth. Regeneration of the forest is directly related to the **life histories** of the tree species, the course of their development from the time they are dispersed as seeds until the time they die.

One controversy that continues to divide ecologists is whether or not natural vegetation is in **equilibrium**, a state in which plants and animals coexist in a stable community over a long period of time. Some ecologists assert that, after disturbance by humans or natural causes, vegetation begins to regenerate and after a number of years reaches a stage of **climax**, its natural equilibrium. This concept is based on the view that an ecological community is like an organism, in that the combination of many separate but interrelated units (here species) function as an integrated whole. Related to this theory is the idea that some communities are closed (they do not permit the invasion of new species) while other communities are open (allowing new populations of plants and animals to become established). An increasing number of researchers believe that there is in fact no state of natural equilibrium, because the species composition and structure of almost all vegetation types are constantly affected by different kinds of disturbance, including that caused by people.

Local people have a detailed understanding of the various stages of vegetational growth in the local environment. This knowledge guides decisions about where and when to cut and burn forest, how many years to cultivate a specific plot of land, where to gather uncultivated plants and where to look for game animals. Each successional stage is identified by observing the height and composition of the vegetation, by recalling the number of years the plot has been left untended and by gauging the extent to which it was originally perturbed. Several characteristics of local perception of ecological succession are exemplified in the Mixe classification of successional stages, presented in Box 5.1.

Agriculturists often intentionally alter the course of ecological succession. For example, abandoned fields are frequently maintained in early stages of succession, because they contain many useful species of plants and often attract game animals out into the open where they can be easily hunted. Pastoralists, who make a living

---

Box 5.1 Mixe classification of the stages of ecological succession

The Mixe recognize several stages of ecological succession, illustrated in Figure 5.4. Primary forest is called **yukj<u>oo</u>tm** and the oldest primary forest, composed of tall and wide-girthed trees, is called **maj_yukj<u>oo</u>tm**.

When forest cover is slashed-and-burned, the resulting field is called **yu'u** *slashed-and-burned field*, referred to as a *rozo* in Spanish. As discussed above, a cultivated field is called **kam** and is further distinguished by the type of crop that is planted.

After several years of cultivation – varying according to the fertility of the soil – a field is left to rest. Weedy herbs begin to colonize the site and these fallow fields are called **kam_tajk** *fallow field*. After a few years, when fast-growing shrubs begin to cover the sites, they are referred to as **peji kam** *shrubland*. Eventually, the area reverts to secondary forest or **aa'my** *secondary forest*.

The Mixe are uncertain whether this secondary forest will eventually return to primary forest, principally because most disturbed sites are maintained in early stages of succession or are reconverted to agricultural fields. **kam_tajk** may be replowed, while **peji kam** and **aa'my** can be again cut and burned in preparation for renewed cultivation.

---

from raising livestock, are often aware of the changes in the species composition of pastures, savannahs and associated woodlands. Through techniques such as burning, they seek to maintain a high percentage of palatable plants while reducing those that are poisonous or too coarse to be eaten.

Indigenous people realize that a forest can regenerate in many different ways, resulting in vegetation types of strikingly different physiognomy and species composition. They may express uncertainty about whether or not regeneration eventually leads to the primary forest becoming re-established, because they have not had sufficient time to observe a complete cycle of succession.

When you set out to document the local understanding of ecological succession, first step back and observe the changes in vegetation from afar and then step into the fields and forests to record the species and structure of each stage. On a walk through the forests of Mount Kinabalu, for example, I observed the stages of succession from a rice field that looks out over villages, fields, forests and the mountain. A farmer from the town of Kiau and a park naturalist from Bundu Tuhan pointed out the transformation from cultivated rice fields (**padi**) to regenerating forests (**temulek**) and primary forest (**puru**). Before continuing on towards the primary forest, I took photographs of this panorama to be used in refining this first glimpse of succession and in comparing how the various stages are classified in neighboring communities.

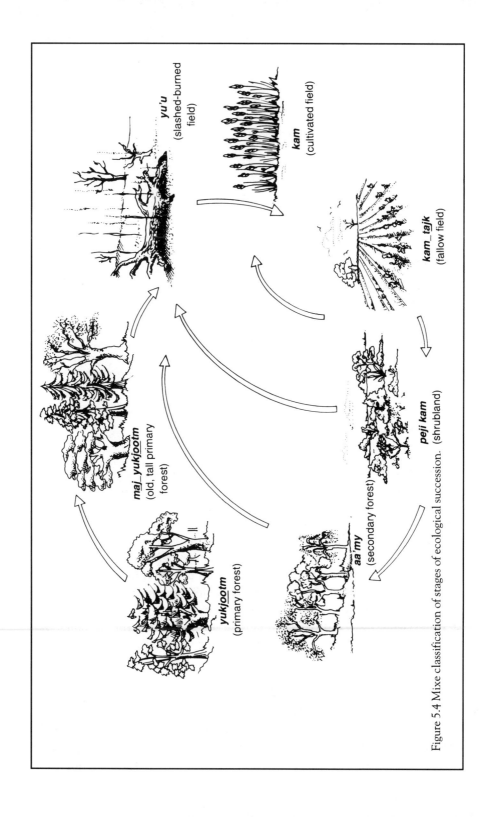

Figure 5.4 Mixe classification of stages of ecological succession.

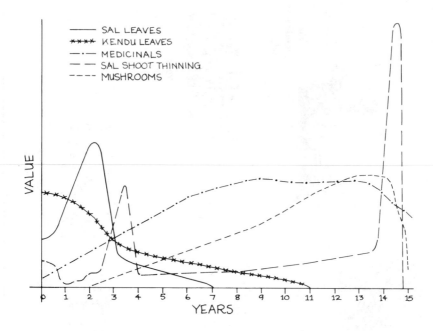

Figure 5.5 Relative availability of forest products in regenerating sal forest, shown as the projected total value of each resource over time.

After you have defined the successional stages from a distance, you can choose a series of sites for closer inspection. A group of researchers in India, including K.C. Malhotra, Mark Poffenberger, Arunabha Bhattacharya and Debal Dev, carried out a rapid appraisal of natural forest regeneration in West Midnapore District in Southwest Bengal [1, 79]. They selected a site in which **sal** (*Shorea robusta*) forest had been recently disturbed and other sites which had been protected for three, five or twenty years. In each plot, they measured the plants along a 10-m line, allowing the researchers and local participants to visualize the relative availability of various resources in the regenerating forests over a span of 20 years (Figure 5.5). Furthermore, they were able to construct the diagram shown in Figure 5.6, which shows how both the species composition and the structure of the forest changes over this period of time.

5.2.6 Land use

As a matter of efficiency, we tend to study one domain of folk knowledge at a time. Yet we realize that each domain is intertwined with the others. As topography changes, so does the climate. Climate affects the development of soil and, together, soil and climate have an impact on the composition and the structure of the

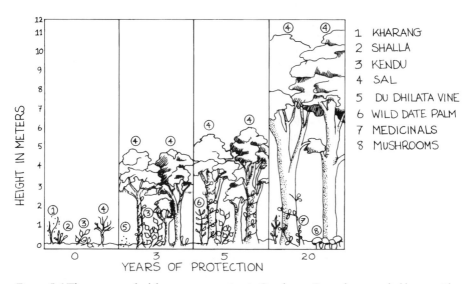

Figure 5.6 The pattern of sal forest regeneration in Southwest Bengal as revealed by a rapid appraisal of ecological succession and availability of forest products carried out in West Midnapore District.

vegetation. In turn, the vegetation alters the development and fertility of the soil and moderates the climate. How can we weave together these various threads and come up with a tapestry as complex as local people's knowledge of the natural environment?

One way is to look at how one domain maps onto another. For example, we can inquire which plants grow in which type of soil or what type of vegetation is found in a certain climate zone. A further step is to observe how knowledge is put into practice as people make a living off the land. The distribution of vegetation types, climates and soils affect local people's choice of where to settle, farm and gather wild plants.

This is the rationale behind studying land use by humans. Classification of the human-influenced landscape has been based on units that are variously called human ecological zones, productive zones, anthropogenic zones or land use types.

Study of human management of the landscape is relatively recent and as yet there is no consensus on how land use units should be categorized. Researchers working in a program called Tropenbos (which means tropical forest in Dutch) suggest classifying land use types based on three dimensions [80]. The first is the **produce** of the land, comprising tangible goods (divided into wood, animals, agricultural crops and forage, as well as extracted fruits, fibers and resins) and intangible benefits, such as protection of biodiversity, maintenance of air and water quality or recreation. The **technology** applied to the land is the second dimension, including human labor, tools, external inputs such as fertilizer and

Table 5.2 A classification of land use types for Indonesia reported by Tropenbos

Permanent cultivation
  Irrigated sawah
  Rainfed sawah
  Tidal rice
  Deepwater floating rice
  Rainfed field crops
  Horticulture
  Homestead garden
  Commercial estate
  Commercial smallholding
  Forest garden
  Intermittent cultivation
  Pastures

Shifting cultivation
  Continuous shifting cultivation
  Discontinuous shifting cultivation
  Recent shifting cultivation

Forested land
  High altitude climatic forest
  Low altitude climatic forest
  Edaphic forest
  Forest plantations

Shrubland
  High altitude shrubland
  Edaphic shrubland

Grassland
  High altitude grassland
  Edaphic grassland

pesticides, as well as landform changes such as irrigation and terracing. Finally, **management** is considered: the size, form and orientation of plots, the intensity of agricultural practices, methods of harvesting, vegetation regeneration, intensity of grazing and burning practices. An example of a land use scheme in Indonesia is given in Table 5.2.

Local people recognize and name land use units in their local environment. These units are at times characterized by their distance from home – the number of hours that must be walked in order to reach a cornfield, collect firewood or gather medicinal plants. Human settlements are usually found at a midpoint in these zones, giving the inhabitants easy access to lands and produce from throughout the territory. Conceived of in this way, the system may be expressed as a series of widening circles – the home is found in the center, surrounded by the community, cultivated fields, ranches and wild areas.

The human ecological zones may also be correlated to categories of soils, climate, vegetation or a combination of these features. Box 5.2 discusses how Chinantec productive zones are related to climatic categories.

Box 5.2 Human ecological zones in Santiago Comaltepec

The Chinantec have access to a broad range of ecological niches in which they carry out a variety of subsistence and commercial activities (Figure 5.7). In ranches located in the hot dry zone, wheat is planted in the autumn months and the rainy summertime often yields a good harvest of beans and corn. A small area of temperate, dry country surrounds Comaltepec, providing a healthy climate in which to reside and one suited to cultivation of subsistence crops and fruit trees. The cold zone divides the humid and dry sides of the municipality and holds reserves of extensive pine forests that yield large quantities of commercial timber. The inhabitants of the humid country dedicate themselves to the cultivation of coffee. In the lowlands, the climate is hot and humid, appropriate for citrus crops, bananas, chilies and cattle-raising.

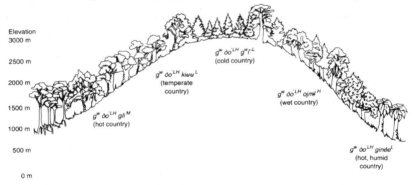

Figure 5.7 A schematic view of Chinantec human ecological zones.

Walking distance	7.5 km	7.5 km	10 km	10 km	10 km
Subsistence activities	Cultivation of corn, beans, squash and chilies; gathering of edible greens, wild fruits, medicinal plants; limited hunting and fishing; firewood collecting; tending home gardens and other activities				
Commercial activities	Cultivation of corn, beans and wheat, formerly cattle raising and cochineal production	Limited poultry and small animal raising; store owners sell diverse plant products	Pine logging	Coffee growing	Cultivation of citrus fruits, bananas, chilies and corn; cattle raising

Different varieties of corn are available for cultivation in these diverse zones, making subsistence cropping a common element in all parts of the municipality. The heavy rainfall on the humid side of the Sierra, spread over 8–10 months, allows for two or three corn crops per year. On the dry side, where the heaviest rain falls over a period of 5–6 months, one annual corn crop is guaranteed in the fields around Comaltepec and, if rains start by June, also in dry hot ranches.

This range of productive zones has encouraged the formation of a flexible economic scheme by which the Chinantec adapt to variable market conditions – a strategy based on the oscillation between subsistence and commercial production and the back-and-forth movement between the climatic zones of the highland Chinantla. When there was a boom in the Zapotec mining towns to the south, the Chinantec increased production and export of wheat, cattle and other goods from the dry tropical zone. When coffee was introduced and promised a good market price, they shifted to the montane cloud forest where the climate was adequate. When external demand for agricultural goods dropped, they fell back on subsistence production in the head town and ranches. From the 1800s up until today, these market conditions guided the Chinantec's management of the natural environment and their use of the land in the diverse ecological zones in their community.

Although you may study these different aspects of the environment one at a time, they are apt to be perceived as integrated parts of a whole to local people. A good way to grasp the interrelatedness of these aspects is to take a walk through the countryside accompanied by men, women and children from the village. Before setting out, discuss among yourselves the dimensions of the environment that are to be observed – productive areas, soils, forest types, landforms and so on. Accordingly, plan out the path that will be taken, ensuring that it passes through a diversity of microhabitats. As you saunter, spend most of your time listening to how people talk about their environs. On occasion, pose questions about the relationships between plants, animals and the non-living elements of the landscape.

If your study area is particularly spacious, think about making several of these walks accompanied by different groups of local people. When you get back, or even while you are still in the field, make a two-dimensional drawing of the lands you have traversed. This can be a smaller scale version of Figure 5.6, which shows the changes in elevation, walking distance as well as observations about productive activities. This can later be turned into a three-dimensional diagram, similar to the one depicted in Figure 5.2, by adding together the results of other transects or by asking people to draw a map of their community.

5.3 Bridging the gap between qualitative and quantitative approaches

In the preceding discussion of qualitative methods, the usefulness of quantifying observations was apparent in several examples: setting up forest plots in Timor to study the species composition of different vegetation types, laying out 10-m lines to document forest regeneration in India, measuring rainfall and temperature in Mexico to understand climate zones.

Quantitative approaches such as these have been extensively used by ethnobotanists throughout the world in recent years. In the following sections, I describe some of the tools, techniques and statistical measures employed by forest ecologists and how they can be applied in ethnobotanical surveys. The basic element in this approach is the delimitation of measured plots in which researchers and local people quantify the number and importance of useful species.

5.4 Quantitative approaches

5.4.1 Choosing an area for study

Making a quantitative census of a forest plot requires much time and effort. Even in a small protected area or community it is an enormous task to survey a small percentage of the total land surface. The solution is to make a careful selection of where to locate the few parcels of land that you will be able to study in detail. Some ethnobotanists use an opportunistic approach, working in permanent plots established for long-term ecological studies. Other researchers set up temporary parcels.

In either case, the plots should be easily accessible and should represent the different ethnoecological zones in the study area. When possible, choose the plots with local people, who can base the selection on their perception of land use, climate, soils, vegetation, successional stages and topography, as outlined in the section on qualitative methods. Villagers know the recent history of land use and can tell the difference between primary and secondary forest better than an ecologist who is unfamiliar with the region.

Alternatively, you can decide to place plots in various types of forest. Oliver Phillips, a British ethnobotanist, worked in six plots established in terra firma, swamp and floodplain forest in southern Peru [3, 4]. Four of the sites were in this last type of vegetation, subcategorized as lower, upper, previous and old floodplain forest. This allowed a comparison of the intensity with which plant resources of the vegetational units are utilized by local people.

Another option is to place the plots at set distances along the ground. You may be interested, for example, in how local knowledge of plants changes as you go 1 km, 5 km or 10 km away from the village center: are the plants most utilized found close to home, with intensity of use decreasing at greater distances? In mountainous areas, plots are often situated at set rises in elevation.

In most ethnobotanical studies, only a few plots are surveyed. There are sophisticated techniques for making a representative selection of sites in a way

155

that is both comprehensive and scientifically rigorous, but also very time-consuming. These methods are used principally when setting up plots which will be monitored over a long periods or where an assessment is needed of the resource potential of a large area that is proposed as an extractive reserve or protected area.

Researchers preselect an area by consulting maps, aerial photographs and satellite images. They then go to the field to verify the exact location and condition of the site, a process referred to as **ground-truthing**. In an area that measures tens or hundreds of hectares, plots or strips of forest are chosen that cover some 3–5% of the total surface area. The sample parcels may be chosen systematically – placing the plots at set distances from each other – or they may be located randomly, using either a simple or a stratified sampling method.

Regardless of how the plots are selected, you should prepare detailed maps and diagrams of how each has been located, showing access trails and natural landmarks. Remember that you or other researchers may want to visit the site in the future to replicate measurements and observe changes in the vegetation.

5.4.2 Measuring the diversity and abundance of plant resources

Researchers often make qualitative assessments of the abundance of a single species or the botanical richness of a forest: 'This forest has a great diversity of plants' or 'This species is common in high mountain forests but is relatively infrequent below 1500 m above sea level' or 'There are three dominant species in this microenvironment'.

In some cases, it is important to quantify these appraisals. For example, you may wish to compare the total number of species found in various types of vegetation or understand the patterns of distribution of a number of economically important trees. If you choose to focus on a specific plant resource, you may want to estimate its overall abundance in the communal lands of one village or to measure the number of individuals found in different size classes as a first step towards understanding the population dynamics of the species.

In recent years, ethnobotanists have taken to addressing these questions by making surveys of useful plants in measured plots. This innovation, inspired in great part by the work of botanists associated with the New York Botanical Garden, has led to numerous comparative studies in the Amazon Basin [81–84]. Other researchers have employed these methods in Africa and Asia. I describe several of these surveys to illustrate the approaches outlined below.

In addition, I draw upon the work of a team of biologists from a biological diversity program sponsored by the Smithsonian Institution and the UNESCO Man and the Biosphere program. These researchers, headed by the ecologist Francisco Dallmeier, are monitoring various types of primary and secondary tropical forest in plots set up in biosphere reserves in Bolivia, Peru, Puerto Rico and the US Virgin Islands as well as other sites in Latin America and the Caribbean [85]. As part of the work in Bolivia, Peru and other Latin American countries, information on local uses and names of plants has been collected. Their research is providing data on floristic diversity and the dynamics of vegetation change in these plots, with the ultimate goal of enhancing the management of biosphere reserves.

These quantitative methods are by necessity participatory – you need a group of three or more people to carry out the work efficiently. Some members of the team measure distances between plants and their sizes, while others make voucher specimens and record data in notebooks. The work can be carried out with researchers who have studied forestry or ecology at a university and local people who know the lay of the land and are familiar with the names and uses of the plants. Tony Cunningham, when studying the harvesting of forest products in the Bwindi Impenetrable Forest of Uganda, worked with teams that included Ugandan park managers, young university graduates and local Bakiga and Batwa people. During a two month period, they made assessments of the use of medicinal plants, basketry materials, bamboo, wood and other resources. The Bakiga and Batwa people provided their knowledge on the location and quality of forest products, while Tony, with the assistance of the park managers and students, lent expertise on methods of resource evaluation.

5.4.3 Quadrats and transects

After deciding where to locate a plot, you must establish its size and shape. Researchers commonly demarcate a quadrat, which is a square or rectangular area used for ecological studies. The method of choosing and delimiting quadrats employed by Francisco Dallmeier and his colleagues is illustrated in Figure 5.8. A 25-ha zone is surveyed and divided into 25 plots of 1 ha each. These plots are then subdivided into 25 **quadrats**, each measuring 20 × 20 m.

These researchers have chosen to define their quadrats as squares of 400 m^2, but plots can be made in other shapes as well. Brian Boom, of the New York Botanical Garden, delimited two plots measuring 20 m by 500 m to study the plants used by Panare people in tall woody savanna and mountain forest in Venezuela. Among the Chacobo of northern Bolivia, he carried out a similar inventory in parcels of 10 × 1000 m (27 141). Plots such as these are referred to as **belt transects**, long and narrow belts of vegetation which may cut across a broad range of microhabitats and species.

The size of the plot is determined by the goals of the study, the number of people in the team and the amount of time available. In order for different plots to be comparable, they should be of the same size and shape. Plots of 1 ha have emerged as a popular standard, used in projects ranging from long-term monitoring of forest dynamics to short-term inventories of useful species. Smaller plots are sufficient for evaluating the quality and quantity of specific resources. For example, Tony Cunningham and his colleagues set up four 10 × 10 m plots in Bwindi Impenetrable Forest to assess the number and quality of bamboo stems per unit area. Jan Salick

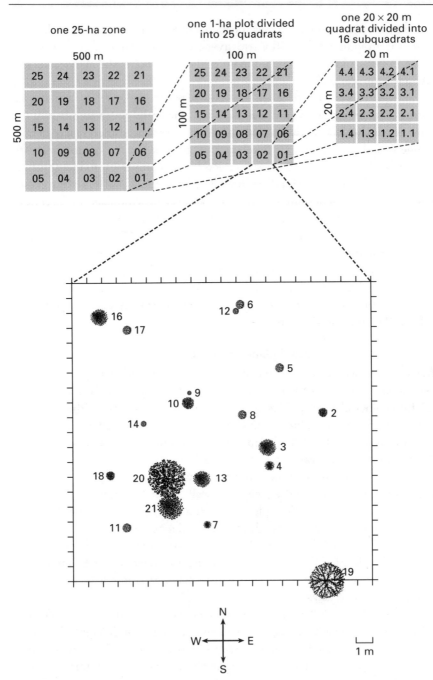

Figure 5.8 The method for selecting and subdividing quadrats employed by Francisco Dallmeier and his colleagues. After the 25-ha zone is divided into 25 1-ha plots, each plot is divided into 25 quadrats (each 20 × 20 m) which are furthered subdivided into 16 subquadrats (each 5 × 5 m).

worked with an Amuesha Indian herbalist in 22 randomly located plots measuring 5 × 5 m to assess the abundance of non-timber resources in the Peruvian Amazon [84].

Once you have decided on the location and size of the plot, how do you go about measuring it? Ecologists use simple tools – meter tapes and stakes – to mark out the transects and quadrats in which they work. A stake is set at a cornerpoint of the square or rectangular plot and the meter tape is used to measure the desired length in the appropriate direction. Because density and other ecological measurements are based on units of flat land, it is important to take into account the slope of the ground so that your results will be comparable with those of other researchers. There are specialized methods and formulas that allow you to calculate the correct length of each side of a plot on sloping land.

Some researchers use string or wire to mark permanently the boundaries of the plot, while others simply use the measuring tape to see which trees fall inside the plot. If you further divide the plot, additional stakes can be placed to mark off the subplots. The quality of the stake used depends on how long you expect to monitor the area. Simple wood posts are easy to manufacture in the field, but they rot in humid forest, burn easily and are a favorite food of some tropical insects. Metal markers, made from aluminum or iron, are durable but more expensive. Because local people use these materials for many other purposes, metal stakes often progressively disappear from the plot. In arid areas with little ground cover, white paint may be used to indicate the corners of a plot but has to be renewed at least yearly.

Describe the quadrat or transect in detail, including notes on the topography, soil and the relative amount of sunlight that filters through gaps in the canopy. Local members of the team can describe the site in terms of their own perception of the environment, employing the qualitative methods described above. They can also provide explanations of how the land has been used in the past. The history of logging, cultivation, resource extraction and enrichment planting will explain much about the present composition and structure of the forest.

5.4.4 Measuring and evaluating individual plants

After the plot is marked out and described, you can begin making a survey of the plants that fall within its boundaries. Which individuals do you choose? Because measuring each plant in a 1-ha plot in tropical forest is an overwhelming task, you need a compromise that allows detection of a large number of species in the plot, while limiting the total number of individual plants that are measured. If your study focuses on a limited number of species, such as the trees used for timber, the individuals that fall into that class are selected. If you are conducting a general ethnobotanical survey, you need to sample a broad range of useful plants.

The criterion used to select individuals of woody plants is usually measurement of the diameter of the trunk at a height of 1.3–1.5 m above ground level (this is referred to as **diameter at breast height** or **dbh**). This is done with a **diameter tape** that gives a direct measurement of diameter when wrapped around the tree trunk. You can also use a standard tape but remember that you must then calculate the diameter by dividing the **circumference** (the distance around the trunk of the tree) by π, a mathematical constant which is equal to 3.142.

Many researchers who work in 1-ha plots select all woody plants which have trunk diameters of 10 cm or more. If you set a higher minimum for the diameter, say 20 or 30 cm, then the number of individuals selected decreases but so does the number of species that will be detected in the plot. Selecting all plants, including herbs, epiphytes and even non-vascular plants can be very time-consuming, but will give a complete species inventory.

If the plots are smaller, then plants with thinner and thinner stems may be chosen without resulting in major increases in the numbers of individual plants in the census. Brad Boyle, a botanist attached to the Missouri Botanical Garden, is comparing the floristic composition, structure and use of cloud forest at three sites in Mexico, Costa Rica and Ecuador. He has chosen to assess plots of 0.1 ha, measuring all woody plants with a diameter of 2.5 cm and above. This approach has some disadvantages. With the smaller cutoff size, he prepares large numbers of voucher specimens from juvenile trees, which are typically sterile, and from vines, which can be difficult to collect. But there are also advantages. With the small plot size, he encountered fewer large trees, which are time-consuming to collect. The number of individuals and species censused is roughly similar in 1 ha/10 cm and 0.1 ha/2.5 cm plots. These smaller plots are particularly appropriate in forests of relatively low diversity such as those which have 100 or fewer species per hectare.

After you have decided the class of individuals that are to be selected, what actual measurements do you make? Diameter at breast height has already been mentioned, but let's take a closer look at how to do it. If the tree is standing straight and has a regular trunk, then it is simply a matter of stretching the tape around the tree and taking the measurement. This is sometimes a job for two people if the trees have very wide trunks. But forest trees are often irregular – insects and fungal attacks create protrusions on the trunk, winds or shallow soil can cause trees to tilt and some trunks just grow crooked or branch at less than a meter off the ground.

Francisco Dallmeier and his team employ practical solutions to this problem, as shown in Figure 5.9. If a tree is tilting to one side, then the measurement is taken at 1.3 m on the side of the tree which is bent towards the ground. On buttressed trees, those with large flanges at the base for support, the diameter is taken at the point where the buttresses end and the trunk begins to narrow more slowly. If there is branching at 1.3 m above the ground, the tape is placed around the trunk at 10 cm higher or lower than the usual height. If the tree or shrub begins to branch very low to ground, then a measurement is made of the two or more stems and the results are added together.

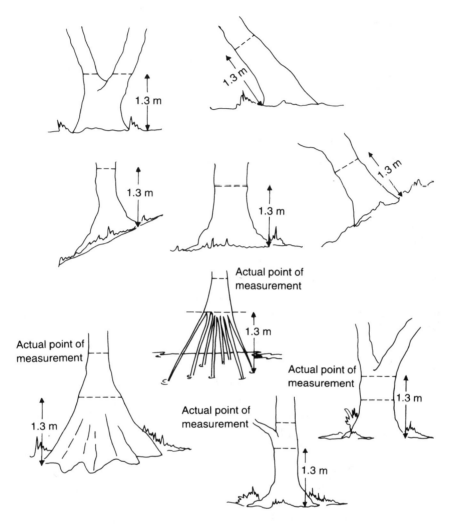

Figure 5.9 Techniques for measuring the diameter at breast height of trees under various circumstances. The center drawing depicts normal measurement of the diameter of a straight, single-trunked tree on level ground. The top four pictures explain how to account for split trunks, leaning trees and unlevel ground. The bottom four illustrations show cases in which the actual point of measurement must be moved up or down to avoid abnormalities in the tree trunk.

Recording the position of the tree is necessary for creating a map of the quadrat which will guide future reassessments of the site. In small plots of 5 × 5 m or even 10 × 10 m, you can judge the approximate position of the tree by sight and sketch it on a map of the quadrat. Although you can estimate the location of individual plants in larger areas, it is more accurate to measure the distance to the tree from

two fixed points, such as the corner stakes of a quadrat, making necessary corrections for slope. This method is used to construct quadrat maps, such as the one in
Figure 5.8, which shows not only the location of the tree but also the relative
diameter of its trunk. Some researchers also indicate the estimated area of the
canopy covered by the crown of the tree, which gives a general idea of how much
space it occupies in the forest.

5.4.5 Estimating the harvest of forest products

After recording the diameter and position of the tree, many researchers estimate
the volume of the plant part which is harvested. When estimating the volume of
commercial timber, foresters gauge the distance from the ground up to where the
tree begins to branch. This span of unbranched trunk is called the **bole**. They
then assess the overall height of the tree from the ground to the top of the canopy.
Tropical trees of 10 cm dbh and above can be very tall, from 20 to 60 m. Many
ecologists just make a visual estimate of the height and this is usually good enough
for ethnobotanical research as well. With a good eye, this sighting technique is
more efficient and may be no less accurate than attempts to measure the tree
directly. In open forest, a tool called a clinometer gives an accurate reading when
used skilfully. However, accuracy with a **clinometer** depends on being able to
stand some tens of meters from the base of the tree, so it is of limited use in dense
forest.

There are no standard ways of estimating the quantity of non-timber forest
products, so researchers improvise techniques for each type of resource. Whenever
possible, accompany resource users into the field to observe which parts are
gathered and to gauge the volume of the harvest. By comparing the potential
harvest with that actually gathered by people, you can understand the efficiency
of collecting activities.

Rattan can be measured in a similar way to timber. The length and diameter of
stem is measured, giving a good estimate of usable cane. For fruit-bearing plants,
the yield per individual can be measured at harvest time either using binoculars
to spot fruits on the tree or simply by counting those that fall to the ground. Other
approaches include measuring the thickness and surface area of bark gathered for
medicinal purposes and recording the number and size of bundles of medicinal or
edible herbs harvested from a measured area. These measurements can be converted to weight in kilograms by weighing a subsample of the collected material
and then multiplying by the total potential amount that could be gathered per
individual or per unit area.

While some members of the team are taking these measurements, others can
make a voucher specimen of the tree and record information on its use and local
classification. The chapters on anthropology, botany and linguistics give full
explanations of how to do this, but keep in mind a few points. Many of the plants

encountered in a plot are sterile and will not yield satisfactory voucher specimens. Fruits and flowers of many of the woody plants will be out of reach, rendering them difficult to collect. If a member of the team can accurately identify the tree by sight, you can avoid making voucher specimens, allowing the data collection to proceed efficiently. But keep in mind that a single common name may correspond to a large number of botanical species, making it difficult to ensure an accurate identification. A local member of the team can return on another occasion to collect fertile material if there is any doubt about the identity of individual trees.

From the cultural point of view, remember that you will be in the field with only a small group of local people. Although they may have a vast knowledge of the forest species, it cannot be assumed that their expertise is representative of other people in the region. Consider returning to the plot a number of different times with various residents of nearby communities or carrying voucher specimens house-to-house to broaden the ethnobotanical information.

Two additional steps will allow you or another researcher to return to the site in later years to assess the growth of the trees and the impact of utilization by local people. First, number the trees so that they can be easily re-located. Second, paint a line on the bark where you placed the tape so that you will be able to make accurate remeasurements of the tree. An inexpensive way of numbering is to paint the figures in the tree, removing some bark if necessary. A much more durable method is to attach an aluminum tag that has the numbers impressed into its surface. These tags can be affixed to the tree with aluminum nails, leaving enough room for the tree to grow without engulfing the tag. If you are monitoring resource harvesting and do not want visibly displayed numbers to affect people's normal patterns of gathering, place the tag on a wire tied to the base of the tree. It may be harder to relocate these trees, but the numbers will be more discreet.

The system of numbering depends on how the plots have been selected and subdivided. At the minimum, one part of the number should designate the plot and another the individual plant. Francisco Dallmeier uses a four-part number: the first two digits correspond to the 25-ha zone or plant community that has been demarcated, the second two digits indicate the 1-ha plot, the third set designate the 20 × 20 m quadrat and the final pair identify the individual plant. In the field notebook, this number should be carefully cross-referenced with the collection number of any botanical voucher specimen taken from the individual plant.

So far in this section, I have discussed the measurement of woody plants. How can a census of herbs, ferns, grasses and other small plants be carried out? In most cases, you will have to come up with creative solutions by adapting the general principles outlined above for the resources that interest you. There will be no trunk diameters to measure but you may be able to estimate the ground surface area covered by a colony of a single species. Alternatively, you could count the number of clumps or stems per unit area. Whatever technique you adopt, keep in mind that it should ensure that a representative sample of plants is surveyed within a reasonable amount of time and effort.

5.4.6 Analyzing the results: density, dominance, frequency and value

Measurements and observations of trees in standardized plots allow you to make calculations that are widely employed by forest ecologists. Because they were originally conceived to measure species richness and the quantity of harvestable timber in a forest, they are easily adapted to ethnobotanical studies.

In order to illustrate how these simple formulae are used, I refer to an ecological study of Luquillo Biosphere Reserve in Puerto Rico. The Bisley biodiversity plot was set up in 1988 as part of the Smithsonian/MAB Biological Diversity program. Not long after the first measurements were taken, a major hurricane passed over the site. It has been re-censused twice since 1989, giving a privileged perspective on the damage caused by major storms.

The first result obtained in ecological studies is the number of individuals and species that are found in each plot. In the original assessment of a 1-ha Bisley plot, 434 individual trees of 10 cm dbh and above were encountered, corresponding to 33 botanical species. This is a relatively low diversity of trees, but not surprising because the area is on an island and consists of secondary forest highly disturbed not only by storms but also activities such as coffee growing, charcoal production and cultivation of subsistence crops.

In other parts of the world, 1-ha surveys of trees greater than 10 cm dbh have given a comparative measure of species richness in various tropical forests. Alwyn Gentry, an American botanist, discovered 300 species per hectare in the Peruvian Amazon, equal to or greater than the richness of lowland evergreen forests in Africa and Asia. In primary montane forests of tropical regions, it is common to find fewer than 100 species/ha. On higher slopes of mountains, diversity drops as elevation increases. In Bwindi Impenetrable Forest of Uganda, for example, Peter Howard found 45–50 tree species/ha between 2000 and 2200 m above sea level and only 20 tree species/ha at 2400 m [86].

Species richness can be further documented by calculating for each species relative density, which is the number of individuals of one species divided by the total number of individuals of all species, multiplied by 100. In the Bisley plot, eight individuals of a fast growing tree of secondary forest, *Cecropia scherberiana*, were found in the 1-ha plot. As noted above, a total of 434 individuals of all species was encountered. This gives a relative density for the *Cecropia* of $8 \div 434 \times 100 = 1.84$.

The relative density gives an idea of the total number of individuals of one species in a plot but does not demonstrate patterns of distribution. Are all the trees of one species clumped together in one spot or are they spread evenly throughout the forest? One way of answering this question is to calculate the **frequency**, which refers to the number of quadrats in which a species is found out of the total number

of quadrats examined overall. For example, in the Bisley plot, *Cecropia* occurred in six out of 25 quadrats within the 1-ha total plot. The sum total frequency of all species was 189 in the 25 quadrats. The **relative frequency** can be expressed as $6 \div 189 \times 100 = 3.18$.

Now we have a handle on how many individuals exist and how they are distributed, but can we say how much space they take up in the forest? From the diameter of each tree, the **basal area** – or area occupied at breast height – can be estimated using a formula which you may remember from geometry: the area of a circle can be calculated by dividing the diameter by 2, squaring it and multiplying by 3.14. If a tree has a dbh of 0.33 m, then the basal area of that tree is $(0.33 \div 2)^2 \times 3.14 = 0.085$ m^2. Usually, basal areas are calculated for each species as a proportion of the total sample area. Thus, the eight individuals of *Cecropia* in the Bisley plot had a combined basal area of 0.94 m^2/ha in 1988. Forest ecologists use the basal area to calculate a comparative measure called **relative dominance**, which is the combined basal areas of all individuals of one species divided by the total basal areas of all species, multiplied by 100. In the case of the Bisley plot, the combined basal area of all species was 28.23 m^2/ha. The relative dominance for *Cecropia* is $0.94 \div 28.23 \times 100 = 3.33$.

While these statistics are familiar to all forest ecologists, there are many other ways to analyze plot data. In particular, ethnobotanists have sought to estimate the value of forest resources using various methods. They may go to a market to ask the price of a certain fruit and then calculate how many fruits could be harvested from the individuals found in a 1-ha plot. This figure is then extrapolated across the entire geographical range of the species. These approaches to measuring monetary values will be further explored in Chapter 6.

Value can also be measured in terms of the significance attributed to the resource by local people. Ghillean Prance and his collaborators calculated use-value of the species in a series of 1-ha plots in two ways: (1) by designating as 'major' or 'minor' the plants' cultural importance and (2) by dividing the plants into a number of use categories, including food, construction, technology, medicinal, commerce and other [83]. Major uses were given an arbitrary value of 1.0 and minor uses a value of 0.5. By summing the use-value for all species by botanical family, they were able to discover which plant groups were of greatest importance to various ethnic groups in Amazonia.

Arbitrary numerical scales may give false impressions of relative cultural value. It would be unwarranted, for example, to conclude that a plant with a combined use-value of 5 is twice as important as one with a use-value of 2.5 or that two plants with use-values of 2.5 are perceived as equally important by local people. These arbitrary scales also introduce distortion in the data because they reflect to some degree the researcher's judgement of cultural importance, based on his or her evaluation of local perception. Another way to gauge relative values is to use the ranking techniques discussed in Chapter 4.

Box 5.3 A formula for estimating use-value based on local perceptions

Oliver Phillips and his colleagues worked in plots of six types of forests in the *Zona Reservada Tambopata* in southern Peru. They studied the uses of many hundred trees and vines of 10 cm dbh or more in a total area of 6.1 ha. The data were derived from 29 *mestizo* (mixed Spanish–Indian descent) people who were interviewed in the forest plots or in their communities. For the purpose of analyzing the results, each act of interviewing a local person on one day about the local names and uses of one species was classified as an 'event'. If a species was encountered more than once in a single day, the person's responses were combined. During 12 months of fieldwork spread over five years, the researchers and local people participated in 1885 independent events.

Oliver Phillips designed the following statistic to analyze the results of this project:

$$UV_{is} = \frac{\Sigma \ U_{is}}{n_{is}}$$

UV_{is} stands for the use-value (UV) attributed to a particular species (s) by one informant (i). This value is calculated by first summing (indicated by the symbol Σ) all of the uses mentioned in each event by the informant (U_{is}) and dividing by the total number of events in which that informant gave information on the species (n_{is}).

Say that you go to a forest in Ecuador with a villager named Juan Torres. Stopping in front of a tree he knows, Juan says the wood is used for construction, the raw fruits as food and the leaves for healing skin wounds. That afternoon, you again find the same species. He repeats his previous observations and adds that the bark is used as an aphrodisiac by some people. The number of uses he mentions in this event is 4. On two other field days, you re-encounter this common species with Juan. One day he says only that the fruits are eaten raw as an emergency food. On another day, he repeats all the previously stated uses and adds that the leaves are used in a preparation to counteract the effects of witchcraft. The use-value for this species in the eyes of Juan Torres is given by the formula stated above:

$$UV_{is} = \frac{4 + 1 + 5}{4} = 2.5$$

This result can be added to use-values derived from other local people ($\Sigma_i \ UV_{is}$). This is then divided by the total number of people interviewed about that particular species n_s to yield the overall use-value (UV_s), as indicated in the following formula:

$$UV_s = \frac{\Sigma_i \, UV_{is}}{n_s}$$

Although this statistic was used initially on results from interviews that took place in forest plots, it could be applied to any data-gathering technique in which numerous people give information on a range of plant resources. For example, if you work with local collectors who make a large number of ethnobotanical collections, each voucher specimen with its accompanying data sheet could be considered an 'event'. It is likely that each species will be encountered numerous times by each collector, so the number of uses on each data sheet can be added together to obtain UV_{is}, the individual use-value. These can be summed for all collectors to calculate the overall use-value for a particular species (UV_s).

The same can be said for house-to-house interviews with fresh or dried plant specimens. You can purposely return with the same species on different days to verify ethnobotanical information. Each time that someone gives information on a species would be considered an event, and the calculations can proceed as outlined above. Once you have the results in hand, you can test a wide ranges of hypotheses. For example, Oliver Phillips and Al Gentry used the *Tambopata* data to compare the non-market value of botanical species and families, the similarity of knowledge between informants and the relationship between people's age and the extent of their knowledge about plants.

Another approach to use-value has been proposed by Oliver Phillips and his colleagues, who worked in the Peruvian Amazon [3, 4]. As further explained in Box 5.3, he bases his estimates of cultural importance on an **informant-indexing technique** which is correlated to local people's agreement on the utility of various species.

Although more sophisticated than other use-values that have been proposed, the informant-indexing technique will need to be tested at other field sites to see if it yields reliable results. Calculating the use-value of biological resources will continue to be a controversial aspect of ethnobotanical methodology.

If you choose to use this or similar techniques, keep several points in mind. First, not all uses are equally important. Consider a species which provides edible tubers used every day and another species which is used occasionally as a source of dye. Can such plants be compared solely on the number of uses they have? This could provide a distorted view of how people assess the importance of biological resources. For example, a plant with five minor uses could be considered more significant to local people than a plant with one major use. A tempting solution would be to rank the uses according to their importance in local subsistence but this would again risk introducing researcher's bias into the analysis.

A counter-argument is that culturally significant plants typically have multiple uses. In Mesoamerica, corn is important not only as a source of food but also because it is used as fodder, medicine and in many other ways. Its high cultural significance would be revealed in an approach based on multiplicity of uses. Oliver Phillips discovered that the palms had the highest ranking of any botanical family in his Peruvian study, something which agrees with the conclusions of other ethnobotanical studies carried out in the Amazon. We know that both palms and corn are both culturally significant and useful in a variety of ways. But other culturally important plants with single uses would be under-valued by this technique.

The number of uses listed may vary according to the context of the interview, the quality of the interaction between the researcher and local people and even the talkativeness of different individuals. When an ethnobotanist first arrives in the community, villagers may be less open to discussing their knowledge about the plant world. As friendships develop and communication becomes more fluent, people open up and begin offering additional information. It is probable that interviews that take place towards the end of the fieldwork will reveal more uses than those held at the beginning.

Even after good rapport is established, the number of uses cited may vary according to the context of the interview. If you have been discussing herbal medicine, it is likely that many curative uses will be cited. Conversely, if you are helping people to forage, the various ways of eating a plant will come to people's minds.

The best way to avoid this source of distortion is to conduct each interaction in a similar fashion. For example, you can emphasize that you are trying to record all uses for each species before starting an identification task. As you come upon each new species in the field, remind people that you are seeking a complete list of uses. As noted in the anthropology chapter, this structured approach to data collection should only be used after you have explored local ecological knowledge through open-ended and participatory approaches.

5.4.7 Estimating total quantities of botanical resources

Once you have measured the density, frequency, dominance and value of the resources in measured quadrats or transects, you can project the overall availability and importance of a resource within a community, protected zone or vegetation type. The precision of the estimate will increase with the number of plots you sample and the accuracy with which you have mapped the various microenvironments of the study area.

In recent years, researchers have developed a sophisticated way of judging the quantity and quality of vegetation by using computers to analyze satellite images and aerial photographs as well as the results of ground-truthing and other data collecting methods. Called **Geographical Information Systems (GIS)**, this technology has the potential to show changes in vegetation caused by habitat destruc-tion and to illustrate where different forest resources are found, giving an estimate of their overall distribution and abundance. Although some ethnobotanists are integrating their data into geographical information systems, other researchers feel that similar outcomes can be obtained at a much lower cost by using manual methods.

Whether by manual or computer-assisted methods, the results of ethnobotani-cal studies from various sites can be combined to give estimates of the total numbers of useful plants that are present in a certain type of vegetation or in a given country. Victor Toledo and several other Mexican researchers are using this approach to calculate the total number of useful products that can be harvested from the tropical lowland forests of Mexico [87]. In addition to the results of surveys they carry out themselves, they review the literature to discover additional uses of the forest by any of the 18 indigenous groups who live in the humid lowlands of Mexico.

By cross-referencing an ethnobotanical database that contains information on 11 of these groups with a database on the floristic composition of the tropical forest, they have predicted the number of useful products that could be harvested from different sized areas, ranging from 1-ha plots to entire regions and to all lowland tropical forests of Mexico. Thus far, they have recorded 1923 useful products that can be harvested from 1024 useful species in the more than 20 million hectares of tropical moist climate in Mexico. Most of these products (69.7%) come from secondary formations rather than primary forests. The majority of the uses (83%) are derived from non-woody parts of the plant, which counters the notion that tropical forests are primarily important as sources of timber.

5.4.8 Surveys of plants grown in home gardens

Home gardens are the usually small plots of land next to a house which are cultivated relatively intensively by its residents. A source of edible, medicinal, ornamental and other useful plants, these gardens are a rich source of ethnobotani-cal information in agricultural communities throughout the world. Many studies of dooryard gardens have been qualitative, consisting primarily of a list of species that are cultivated or managed by the home owner. Yet these sites are ideal for carrying out a quantitative assessment using ecological techniques.

There are some special characteristics of gardens that require modification of the techniques used in forests. The sample of plots is based on a selection of households to be visited and whether or not the individual homeowners agree to participate. The size and shape of each plot is predetermined by the boundaries of the home garden and this varies from one household to the next. Almost all the plants found within the plot are culturally important, which means that not only trees and shrubs but also herbs and grasses are included in the study.

Apart from these differences, a plot survey is carried out in much the same way as described above. A map is drawn of the garden, indicating the dimensions of the plot and the position of plants, the house and any walls or fences. The number of individual plants is counted and the species of each is determined.

A voucher specimen may be collected, but there are good reasons to avoid this in garden studies whenever possible. The plants are usually common and easily identified on sight. There are often few individuals of any one species in each garden and these could quickly disappear, roots and all, in the hands of an eager ethnobotanist. The residents of the house depend on the garden for everyday subsistence and they would prefer to see the harvest end up in a pot rather than a plant press.

Once you have recorded the size and composition of each garden you can analyze the data using the statistics presented above. The results may reveal great variation between sites, as can be seen in a study carried out by a group of Mexican researchers, headed by Victor Rico-Gray [88]. They made a survey of home gardens in two Maya communities, Tixpeual and Tixcacaltuyub, in the Yucatan Peninsula. The 20 plots visited in Tixpeual varied from 400–5000 m^2 whereas the gardens in Tixcacaltuyub measured between 800 and 3200 m^2. In both cases, these sites added up to a total of over 40 000 m^2, providing a good basis for comparison of the two communities. More than 5600 individual plants and over 130 species were de-tected in each village. Despite these superficial similarities, there was great variation in several statistical measures, including the species similarity within and between the villages, and the correlation between the size of the plot and the number of individuals and species contained.

Along with this quantitative approach, you may also employ a participatory technique to understand the layout of each garden. By drawing a map with the owner of each house, you can discover if there is a logic behind the placement of individual plants or whether this is random. At first, you might be fooled by the disorganized aspect of the garden, but look for patterns that can be explained by social, cultural or geographical features. Differences in home gardens within a single community are often correlated with ethnicity, relative wealth and family organization of the owners.

When comparing gardens between villages, note differences in climate and topography as well as the relative distance from urban areas. Victor Rico-Gray and his colleagues noticed that the village gardens in the community closest to the city of Merida contained a relatively high number of ornamental plants when com-pared to the plots of the more distant community. This may indicate a general change in the function of home gardens, which are perhaps more linked to subsistence in remote villages and serve to supply city markets or play an aesthetic role in urbanized areas.

6

Economics

Figure 6.1 A temporary market stall in the *Mercado de Abastos*, an urban market of Oaxaca City. Marketplaces provide opportunities to survey the diversity and monetary value of economically important species of plants and animals.

6.1 Economics and ethnobotany

Although business and trade have existed for thousands of years, it was only in the 18th century that a group of European scholars began to formalize the discipline we now refer to as economics. Economic evaluation has always played an important role in studies of how plants are used by local people. Some researchers refer to their work as **economic botany**, putting the emphasis on the discovery of plant resources that attain importance in global or regional markets, thus possibly contributing to national and community development. At present, researchers are making a concerted effort to assess the value of non-cultivated resources that are harvested from forests and fields [89–95]. This forms part of an effort to demonstrate the economic benefits of conserving forests and documenting traditional ecological knowledge [5, 96-98].

6.2 The value of the environment

Before exploring various techniques for assessing the value of biological resources, it is important to look at some of the ways that economists characterize natural resources. They contrast **non-renewable resources** – substances which exist in limited quantity and are potentially exhaustible – with those that are **renewable**, substances potentially available indefinitely. These concepts are linked to the timescale in which people live. Non-renewable natural resources like coal and oil take aeons to form, whereas renewable resources can be reproduced within a few years or generations. Although plants and animals, and products derived from them, are often considered as renewable resources, the ability to maintain them depends in great measure on how they are managed. Over-exploitation can leave certain species without commercial value or can even drive them to extinction.

In recent years, environmentalists and economists have been attempting to describe the different components of value derived from the natural environment [99]. Paul and Mary Ehrlich, American ecologists known for their popular books about the environment, divide the value that people gain from nature into four categories [100]. **Direct economic value** is derived from the food, construction materials, medicinal plants and other goods harvested from natural areas, as well as from recreational facilities and other services. **Indirect economic value** refers to benefits provided by various other environmental services, including regulating climate, purifying water, maintaining the fertility of soils and absorbing waste products. **Aesthetic value**, based on the appreciation of the beauty of nature, is manifested when people engage in ecological tourism, including birdwatching, nature photography, scuba diving and hiking. It may be also derived from the more abstract pleasure of knowing that natural areas, plants and wild animals are available even if a person never travels to see them. **Ethical value** comes from the belief, common to many cultures, that life has intrinsic significance – its value is derived simply from the fact that it exists. The overall significance of natural areas, for example, based on these four categories, is potentially immense, but is rarely taken into account by traditional economists, who typically base their estimates only on direct economic value.

6.3 The value of forest products

Demonstrating economic value can proceed by making an independent assessment of each of the components of value. Assessing the direct use value of forest resources does not preclude making a later evaluation of their indirect use, and aesthetic or ethical values. Later, these different assessments of economic worth can be evaluated to gauge the total economic value of forests. The process of calculating overall value is relatively new in economics, so the methods are still experimental and not well established or universally accepted. Most ethnobotanists are far from being able to assess all these components of value and they are even novices at analyzing the economics of forest products.

Miguel Pinedo-Vasquez, Daniel Zarin and Peter Jipp, all from the School of Forestry and Environmental Studies at Yale University in the United States, have discussed some of the limitations of current methods and sources of data in their analysis of the economy of a community in the Peruvian Amazon [95]:

> A complete analysis of the economic advantages of different land-use alternatives would require full knowledge about the entire range of benefits, market and non-market, accruing from each option. Production and price data for many fruit species and for all forest-dwelling game species are currently unavailable. Efforts to assess the value of forest products in subsistence economies are also lacking. Furthermore, there are no regional data assessing the value of ecosystem services of intact forest, such as watershed protection.

They suggest that while this complete data set is missing, the best way to measure the relative profitability and attractiveness of various land-use alternatives is to examine the economic opportunities that local people have at their disposal. In the remainder of this chapter, we will look at various interrelated methods of assessing the current and potential value of forest products by carrying out economic analyses of forest plots, households, communities and markets.

6.3.1 Calculating the value of forest products in measured plots

One way of assessing the value of a forest is by comparing the **rates of return** or potential financial benefits derived from different ways of employing capital and labor. We can ask which way of exploiting forests would earn the greatest income from each unit of land, a measure that economists call the **rate of return to land**. For example, would it be more profitable to convert a forest to cattle pasture, turn it into a timber plantation or leave it standing to allow harvesting of non-timber products? This type of comparative economic analysis measures the direct use-value that can be derived from these various production strategies, an approach which economists call the **change of productivity method** [99].

Analyses of the profitability of commercial agriculture, timber operations and cattle raising are relatively easy to find in the literature. Although studies of forest resource extraction are rarer, several recent case studies – such as those described in Boxes 6.1 and 6.2 – reveal various techniques that can be used to assess the value of non-timber forest products.

The first step is to make an inventory of all the useful organisms in a certain type of vegetation. As discussed in the chapter on ecology, this is usually done by demarcating a plot and documenting the useful plants within its boundaries, an approach which allows the results to be projected over a much wider area. Although most researchers measure just the trees and vines that have a diameter of 10 cm and above, the method can be modified to include the many herbs, ferns, mushrooms, shrubs and other plants that have medicinal, ornamental and other uses. Other techniques can be used to measure the quantity of animals available in the forest, giving a complete understanding of the breadth of resources that have current or potential market value.

After discovering what resources are available, you can proceed to measure the potential harvest that can be reaped from each species. As noted in the ecology chapter, the most precise method – direct counting or weighing of the fruits, bark, foliage or other marketable material produced each year – is also very time-consuming. The measurements from a number of individual plants or populations may be averaged and then multiplied by the number of mature individuals in the plot.

Another way to estimate the amount of harvestable products available in the forest is to talk with gatherers. Because they depend on forest products for both subsistence and income, they typically have a detailed knowledge of the dependability and quantity of the yields of different species. In order to cross-check the responses, you may wish to create a seasonal diagram of the relative availability of various resources throughout the year. You can also use ranking techniques, as described in the anthropology chapter, to verify the relative quantities of each product harvested.

Whether you measure yield directly or by interviewing local gatherers, keep in mind that the yield can vary greatly from one year to another. For example, during a good year the harvest of the edible seeds from the Chalghoza pine (*Pinus gerardiana*) provides income for about 13 000 people in indigenous communities of the Suleiman mountains of Pakistan. During normal years, many fewer people can make a living from this resource. Dipterocarp forests in Southeast Asia show similar annual fluctuations, with heavy **mast** fruiting followed by several years of low production. Enquiring about this variation and averaging the harvest over a several-year cycle permits a more accurate measurement of yield in these forests.

Oliver Phillips, who carried out a detailed study of the potential fruit harvest in tropical rainforests of the Peruvian Amazon, recommends that researchers pay attention to several other factors that affect the overall yield that is accessible to local people. The productivity may vary from one forest type to another [2]. He found that seasonal swamp forest yielded many more fruits and nuts per hectare than terra firma forest, which is the most common vegetation type in the Amazon. The production may be clustered in one season of the year. In the Amazon, for example, fruits were produced in greatest abundance in the wet season months between November and April. The ease of harvesting fruits and nuts varies with the species. Fruits which fall to the ground or are produced not far above breast height are more accessible than those which emerge and remain near the canopy. It is only by taking into account these sources of variability that we can make an accurate assessment of the total amount that is harvestable in an average year.

Once you have determined the total potential yield of each species, make an estimate of the amount that can be gathered each year without affecting the regeneration or future yield of the species. Leaving some amount of forest products for regeneration is an accepted principle of resource management, but the exact percentage that must be left to permit a maximum sustainable yield is difficult to calculate and varies from one species to another. A precise calculation of the maximum level of sustainable harvesting would require monitoring the population over a number of years, a technique which requires much time and work. Most researchers choose an arbitrary level for all species in order to simplify the estimation of net present value. In the Peruvian case discussed in Box 6.1, the researchers assumed that leaving 25% of the fruit harvest for forest regeneration would be sufficient to ensure the viability of the various species.

Box 6.1 Putting a price on forest products in Peru

One of the most famous attempts to put a dollar value on non-timber forests products was carried out by a multidisciplinary team composed of an ecologist, a botanist and a resource economist [94]. Charles M. Peters, Alwyn H. Gentry and Robert O. Mendelsohn based their findings on an inventory of 1 ha of tropical primary forest near the village of Mishana, Peru. They found 50 botanical families, 275 species and 842 trees with a diameter of 10 cm or greater. Of these, 72 species (26.2%) and 350 individuals (41.6%) yield products that have a market value in the nearby city of Iquitos. Sixty species produce timber, 11 species (including four palms) provide edible fruits and one tree species yields rubber.

These researchers calculated the annual production of the fruit and latex-bearing species and measured the volume of commercial wood in each timber tree. Prices for these products were estimated by visiting local marketplaces, sawmill operators and the governmental office that controls rubber prices.

They discovered that the 1 ha of forest at Mishana produces fruit worth almost $650 each year and that the rubber yield is worth about $50 annually.

They deducted labor and transportation costs from these totals to arrive at a net annual revenue of $400 from the fruits and $22 from the rubber.

Using the equation NPV – V/r, with the discount rate r set at 5%, they arrived at a net present value of $6330 per hectare. They further calculated that timber from the plot, when harvested in a way that does not damage the production from fruit and rubber trees, would have a financial worth of $490. This gives a total NPV of $6820. By contrast, the amount of money that could be earned by cutting all the timber in one operation is estimated at a little over $1000. Based on these calculations, Peters, Gentry and Mendelsohn argue that sustainable management of forest would yield greater long-term financial benefits than clear-cutting.

6.3.2 Establishing market value

After making an inventory of useful species and an estimate of their annual sustainable production, you can proceed to establish the market value of each resource. Some products which are important on a national or international level, such as rubber or brazil nuts, have a price which is established by trade in the commodity. In some cases, the national government may set a minimum or maximum value or fix the price at a certain level. For these resources, the current and past prices can be obtained by visiting the governmental agrarian or forestry offices responsible for economic policy on forest products.

For minor forest products which are traded in local marketplaces, the price can be established by making periodic visits to a number of vendors, as discussed towards the end of this chapter. Whatever method you use to establish the current value of a resource, remember that prices of agricultural and forest products fluctuate widely according to supply and demand. You should try to take these temporal changes in market dynamics into account by documenting how the price of a commodity varies over the course of the year and how it has changed in recent history. If you are interested in calculating the returns to the producer, keep in mind that a person who gathers forest products does not necessarily receive the full market price. If you are working backwards from the market price, you should document how prices decrease along the chain of retailers, wholesalers, middlemen and gatherers.

When you multiply the size of the harvest by the market price, you calculate the gross revenue or total earnings that can be obtained from a forest product. Gross revenue is not pure profit because it does not include deductions for the costs involved in the harvesting and processing of forest products, as well as of bringing them to market. The net revenue, which gives a realistic picture of the profits earned from a commodity, is derived by subtracting from the gross revenue all labor and transportation costs.

The first step in calculating labor costs is to measure the total amounts of time spent by the various people involved in harvesting, transportation and selling. Whether through interviews with local gatherers or participant observation in the field, you should determine how long it takes to harvest products in the forest and deliver them to the marketplace or wherever else purchasers congregate. Be aware that some labor may not be evident at first sight, including the work involved in managing the resource – planting seedlings, pruning trees, weeding to remove competitors or applying any other horticultural techniques that are directed towards maintaining current yields. The total number of hours or workdays is multiplied by a rate of pay – usually the minimum wage fixed at the national level or the typical daily wage an agricultural laborer earns in the region – to calculate the total labor costs.

In addition to labor costs, there are other expenses associated with marketing forest products. Harvesting tools must be replaced on occasion. More significantly, transportation implies paying for fuel and maintaining a vehicle, or covering freight costs on buses, trucks, planes or other forms of public transport. There may also be official taxes levied according to the amount of material which is being sold. Vendors in public marketplaces must often pay for the right to occupy a stand.

At times you may be able to measure the exact cost of transportation, for example the amount that it costs to bring a shipment of fruits to a marketplace by boat or by bus. In other cases, you may have to make a rough estimate of transportation costs as a set percentage of the market price of the product. In the Peruvian example, transportation was assumed to cost the equivalent of 30% of the total market value for fruits and rubber latex. The costs of transporting products from places of harvest to points of sale increase with distance and difficulty of access, so this percentage will vary from one place to another.

6.3.3 Calculating net present value

The assessment of total monetary value must take into account not only the current market value of one year's harvest, but also the potential production in future years and alternative opportunities for using the land and investing the profits. In the assessment of minor forest products in the tropical forest of Peru, researchers used a simple equation to estimate the **net present value (NPV)** of minor forest products:

$$NPV = V/r,$$

where V is the net revenue produced each year and r is an inflation-free discount rate of this annual income. For the purpose of this calculation, they set the inflation-free discount rate at 5% and assumed that 25% of the fruit would be left in the forest each year for regeneration.

Because the **discount rate** is a controversial concept, opinions about the correct way of estimating the percentage of discount differ from one researcher to the next.

To understand the discount rate, we must first observe that there are many assets in any economy which can yield income. There exists a choice between conserving an asset to take advantage of its future value or cashing in on it immediately and investing the profits in such a way as to yield further income later. For example, forests can either be conserved for their future value or logged and converted to pastures or plantations in order to gain wealth which can be invested in other money-making opportunities.

To determine the best economic options in terms of income, we have to compare the potential returns from the various possibilities of managing assets, allowing us to determine the general **rate of interest** in the economy. If we focus on nature conservation, for instance, gauging the likely income generated from intensive resource harvesting or forest conversion will allow us to estimate the cost of preserving the forest. The future income that is earned from preserving these wildlands must be discounted by the rate of interest earned on other assets, giving a realistic assessment of the economic benefits of conservation.

The theory is fine, but how do we put it into practice? As in the Peruvian case study, many researchers choose an arbitrary discount rate of 5%. A more prudent approach is to compare the net present value obtained when using a series of different discount rates. For example, Miguel Pinedo-Vasquez and his colleagues recalculated the results presented in Box 6.1 using discount rates of 5%, 10% and 15%. They point out that the high estimate of US$6330 based on a 5% discount rate dwindles to US$3165 when r is set at 10% and to US$2110 when r is 15%. There can be no universal solution for all ethnobotanical studies, because the appropriate discount rate for any particular region will depend on the economic factors at play in the country.

To review the chain of steps used to establish the value of a forest resource, let's take an example from the Peruvian case study given in Box 6.1. In the 1-ha plot, the researchers discovered three individuals of *Parahancornea peruviana* Monach, a tree in the Apocynaceae which is called **naranja podrido** (*rotten orange*) in local Spanish. Through interviews with local collectors, it was determined that each tree produces an average of 150 fruits per year. Monthly surveys of prices in the Iquitos market indicated an average price of 25 American cents per fruit. The total annual value or gross revenue is thus calculated as (3 trees) × (150 fruits per tree) × (25 cents per fruit) = $112.50.

If we estimate that it takes two days to harvest each tree and we use the minimum wage of $2.50 per day, then we would subtract $15 for labor costs ($2.50 per day × 2 days per tree × 3 trees = $15). Transportation costs would be $33.75 ($112.50 of gross revenue × a standard transportation cost of 30% = $33.75). The net revenue, assuming the gatherers bring their own fruits to market, would be $63.75: ($112.50 net revenue) − ($15 labor costs + $33.75 transportation costs). The net present value for *naranja podrido* is then $63.75 ÷ 0.05 = $1275.

Fruits and latex can potentially be gathered year after year, but what about resources that are clear cut, left to regenerate and then reharvested after many years? Calculation of the net present value of this kind of cyclical exploitation requires a different equation, referred to as the Faustman formula:

$$NPV = R/(1 - e^{-rt})$$

where R is the net revenue produced each year, r is an inflation-free discount rate of this annual income and t is the number of years the resource is left to regenerate. e is the base of the natural logarithm (a mathematical constant that equals 2.17828...). An illustration of this calculation is given in Box 6.2.

Box 6.2 Raising revenues from medicinal plants in Belize

A resource economist, Robert Mendelsohn, and an ethnobotanist, Michael Balick, joined forces to assess the value of medicinal plants in Belize, Central America [90]. They note that the most common methods of collecting medicinal plants in Belize are destructive, including felling trees, stripping all the bark from woody plants or digging up the roots of herbs and vines.

Replicating these methods, Mendelsohn and Balick set out to estimate the net present value of all medicinal plants that they could clear from two small plots covered with secondary forest. One plot of 0.28 ha was delimited in 30-year-old valley bottom forest at 200 m above sea level. It yielded a total of 86.4 kg dry weight of medicinal plant parts, harvested from five different species which are sold locally. The second plot measured 0.25 ha and was set in 50-year-old forest on a ridge in the foothills of the Maya mountains at about 350 m in elevation. From this plot, the researchers were able to gather 358.4 kg dry weight of marketable plant medicines that corresponded to four botanical species.

In order to extrapolate how much material could be found in similar plots of 1 ha, they divided the yield by the size of the plot: 86.4 kg ÷ 0.28 ha = 308.6 kg per hectare in the 30-year-old forest and 358.4 kg ÷ 0.25 ha = 1433.6 kg per hectare in the 50-year-old forest. Estimates from local herbal pharmacists, healers and farmers indicate that each kilo of unprocessed medicinal plants could be sold for an equivalent of $2.80 (expressed in American dollars). The gross revenue from medicinal plants is thus calculated by multiplying this price by the estimated yield. One hectare of the 30-year-old forest would bring revenues of 308.6 kg × $2.80/kg = $864, whereas a hectare of 50-year-old forest would yield 1433.6 kg × $2.80/kg = $4014.

Labor costs are calculated by multiplying the local wage rate ($12/day in American dollars) by the number of days required to clear each of the small plots (7 days in the younger forest and 20 days in the older forest) and then extrapolating these results to areas of 1 ha. This gives a harvest cost of $300 per hectare of 30-year forest and $960 per hectare of 50-year forest. Transportation costs are considered to be negligible, so the net revenue per hectare is calculated as $864 – $300 = $564 for the younger forest and $4014 – $960 = $3054 for the older plot.

If the Belizean herbal medicines were being collected in a sustainable way that did not kill the individual plants, the net present value could simply be calculated by the equation NPV = V/r, as in the Peruvian example. But because the medicinal plants are totally cleared from the plots, a period of regeneration before reharvesting must be included in the calculation. Mendelsohn and Balick estimate the time of regeneration (called the **rotation length**) as the current age of the forest, that is, 30 and 50 years for the two different plots. Using the formula NPV = $R/(1 - e^{-rt})$, with a standard 5% discount rate, they calculate a present value of $564/(1 - 2.178^{-0.05 \times 30})$ = $726/ha for the first plot and $3054/(1 - 2.178^{-0.05 \times 50})$ = $3327/ha for the second plot. These figures compare favorably to the revenues generated by alternative land uses in the tropics, including intensive agriculture, subsistence farming and tree plantations.

6.3.4 Calculating the value of non-marketed biological resources

As in the case studies in Belize and Peru presented above, most ethnobotanists base their estimates of the commercial value of wildlands on the direct use-value of marketed renewable resources. Although this is a first step towards understanding the value of the forest, we should not forget that plants and animals which are not marketed also play an important role in the subsistence of local people and may have a future market value which has yet to be discovered.

Rural people consume hundreds of species of edible plants which are never found in marketplace stalls. They often construct their houses with woods that are not sold internationally, or even locally. There are thousands of medicinal plants, known and used by local people, which are not sold in markets. If these plants are nourishing, sheltering and healing a large amount of the world's population they must have economic value, but how do we estimate it?

Economists find it difficult to place a price on resources that have a significant use-value but no current exchange-value. Some would argue that they have no monetary value because there is no current market demand for them. Other researchers propose that their value is infinite and cannot be measured in terms of cash prices. Both positions are unsatisfactory, because they effectively leave non-marketed resources out of the economic valuation of natural areas.

In particular case studies, ethnobotanists often put an arbitrary value on forest products. For example, if most medicinal plants cost $0.50 a bundle in a rural marketplace, this same price will be attributed to medicinal herbs which are not currently marketed. This approach is sufficient when assessing household or community economy, as further discussed below. On a global level, it probably under-estimates the true value of the resources and certainly does not allow for the possibility that local knowledge of some species could lead to the discovery of million-dollar pharmaceuticals.

Economists suggest several other experimental solutions. For example, they might direct their inquiry by asking how much it would cost to substitute for the item, a value which is referred to as the opportunity cost. If a plant which is said to cure headaches were to disappear, how much would people have to pay to obtain an equivalent medicine in the pharmacy? If, instead of using a local medicinal herb, a person travels to a hospital to see a doctor and purchase medicines, how much time, energy and money would be spent? A similar approach is referred to as the willingness-to-pay technique, which consists of asking people how much they would pay in order to maintain their access to a certain good or service.

Although these techniques encourage researchers and policymakers to reflect on the total value of forests, it is impractical to apply them in most ethnobotanical studies. The questions are probably too hypothetical to elicit consistent responses from people in rural communities, who may find it difficult to put a cash value on all goods and services which they obtain through subsistence activities and bartering.

Some researchers have abandoned trying to set monetary values for naturally occurring plants and animals. They prefer to use other measures such as the relative importance of each organism for subsistence. For example, the value of edible plants can be measured by analyzing their nutritional quality and judging their percentage contribution to minimum daily dietary requirements. Another method of estimating the value of edible plants and animals is to measure how much energy is expended in harvesting them compared to the amount of energy they provide when consumed.

Because these approaches require extensive data gathering and laboratory tests, it is unlikely that they will be adopted in most ethnobotanical studies. For the most part, researchers will continue to focus on the direct use-value of marketable resources. Yet it is critical that we do not forget the significance of subsistence activities and non-marketed resources that add to the total value of the environment.

6.3.5 National and regional trade in forest products

The methods discussed so far emphasize the potential value of forest resources per unit of land. Yet as discussed previously, there are many reasons to question the accuracy of these calculations. The fruit yield may be highly variable from year to year and market prices may fluctuate. As a predictor of how people choose to make a living, the rate of return from the land is less reliable than the return on labor. Instead of measuring the potential value of biological resources, you may wish to calculate directly the actual yield and yearly income that countries, communities or households derive from each forest product as well as the actual amount of time dedicated to harvesting.

How can we estimate the actual income earned from forest product gathering during the course of a year? The key is obtaining data through both official records and interaction with local people. You can start by looking at reports from governmental agencies, unions of harvesters and private enterprises. If the government is involved in buying, taxing or regulating the export of biological resources, there may be annual reports on the quantity and value of the overall trade. You will usually find that the data is broken down by region or given for the country as a whole, but in some cases there will be village by village or even family by family accounting. For example, as mentioned in Box 6.4 (p. 189), the Gujarat State Forestry Development Corporation in India keeps track of the non-timber forest products that it acquires from villagers in the region [101].

If private entrepreneurs purchase the products, they may give you access to data on the volume of material processed each year. Plantecam Medicam, a French owned company based in southwest Cameroon, issues annual reports on the amount of *Prunus africana* bark that is brought to its factory. These data allowed Tony Cunningham and Fonki Mbenkum, who are studying the sustainability of the trade, to calculate that 9309 tonnes of bark were officially purchased by the company between 1985 and 1991 [102].

Keep in mind that it can be to the advantage of producers and buyers to variously under-estimate or over-estimate the volume traded. For reasons of pride and prestige, some people may exaggerate the amount of income that they earn on a yearly basis. Others, in an attempt to avoid governmental taxes or the jealousy of neighbors, will minimize their wealth. Where hunting animals and harvesting ornamental plants is illegal, producers and buyers will usually be unwilling to discuss their trade.

On a regional or national level, it is common to under-report the weight and value of forest products in order to escape the payment of some governmental taxes or export tariffs. In many cases, given the relatively minor role of forest products in the economy, official records are kept poorly or not at all. You may find, for example, that all medicinal plants are grouped together under one export item and that no data are available on individual species. For these reasons, you should remain sceptical of official records and look for other ways of verifying the data. Tony Cunningham and Fonki Mbenkum checked another source – the weighbills kept by local entrepreneurs as they buy bark from local harvesters – which showed that 11 537 tonnes of *Prunus* bark had been purchased between 1985 and 1991 by Plantecam Medicam, the sole buyer in Cameroon. By combining information from various sources, they were able to obtain a reliable estimate of the range of material which was being harvested in the country each year.

6.4 Surveys of community and household economy

Besides the possible inaccuracy of official reports, they are rarely available on a household or even community basis. Yet many of the theoretical concepts of economics are brought down to earth only when you look at the productive activities of individuals. Rural people have to decide how to make a living on a daily basis, choosing between options such as timber extraction, slash-and-burn agriculture, harvesting of forest products and earning wages by working in their own community or migrating elsewhere.

Although surveys are often used to gather information on the local economy, people are sometimes hesitant to answer questions about their personal wealth and the productive activities in which they engage. In addition, they may not be accustomed to calculating the total yield of a certain forest plant or the amount of time they spend hunting every year. They may have different perspectives on assessing wealth and estimating the values of local resources. For these reasons, it is often best to start an analysis of household or community economy with techniques drawn from participatory rural appraisal rather than with direct interviews [9].

6.4.1 Wealth ranking

One of the first exercises that local people can carry out is **wealth ranking**, which provides a comparison of the relative income and assets of members of the community. This ranking exercise, which is similar to other types of ranking tasks discussed in Chapter 4, draws upon the differences in household wealth which are linked to social status, land ownership, behavior and productive activities. For example, researchers who work in stratified communities – those in which there are significant differences in the social status or wealth of the residents – often find that gatherers of forest products are among the poorer members.

Apart from revealing the relative position of households in the community, wealth ranking allows researchers to discover the criteria that local people use to measure wealth. The results of this exercise can be used to characterize the segment of the population most dependent on resource harvesting and to select specific families to be interviewed more intensively.

To begin, list all households in the community and assign a number to each. This roster can be drawn from a census that you have carried out previously or simply by asking participants to enumerate all households at the beginning of the exercise. Working from the master list, write the name of each household head and the corresponding number on separate notecards. Invite three or more people who have lived for a long time in the community to sort the name cards into separate piles, each corresponding to a different level or category of wealth. Once all the cards have been sorted, ask the participants to describe the members of each pile, suggesting the essential indicators of wealth or poverty that characterize each group.

This exercise is best carried out in private with one participant at a time because people are often hesitant to discuss sensitive economic issues in public. The cards should be shuffled before each sort in order to ensure that they are presented in a random order.

Although the gross wealth groupings may serve as the basis for further research, you can gain a more detailed ranking by calculating the relative score of each household. On the master list, write next to the appropriate household the number of the pile into which each participant has placed the corresponding name card. The pile of wealthiest people is considered to be #1 and the piles of decreasing wealth are numbered consecutively #2, #3, #4 and so on. If everyone uses the same number of piles then you can simply add the numbers together and divide by the number of people who have participated.

If different numbers of wealth categories are recognized by the various participants, then you will have to use some simple mathematics to make the results comparable. Divide the score of each household by the total number of piles used by the participant and then multiply by 100. For example, for a person who divides the cards into five piles, the score for a household placed in the fourth pile would be 80 ($4/5 \times 100 = 80$). For a person who divides the cards into four piles, the score for a household placed in the third pile would be 75 ($3/4 \times 100 = 75$). By adding these modified scores and dividing by the number of participants, you can obtain a detailed ranking of household wealth.

You can later cross-check the ranking as you get to know people in the community better and carry out more detailed interviews. Keep an eye out for discrepancies between the results of wealth ranking and the indicators that you observe using other methods. In some communities there is a taboo on accumulating wealth, so people may be reluctant to divide community members into distinct groups. In other villages, participants may systematically exaggerate or under-estimate the well-being of the community because they wish to elevate the status or underline the needfulness of villagers. For these reasons, it is difficult to compare rankings between communities.

6.4.2 Livelihood analysis

The ways in which people make a living can be rapidly conceptualized by **livelihood analysis**, which comprises various techniques that allow participants to interpret the behavior and coping strategies of different households. The central element is a data table which contains the names of household heads and the corresponding house number across the top of the page and, along the left-hand side, categories such as sources of income, number of animals owned, estimated size of the most recent harvest and additional indicators that you have discovered through wealth ranking or other approaches. The data can come from a variety of sources, including interviews, participant-observation and analytical tasks that have been carried out previously in the community.

Table 6.1 Example data table for livelihood analysis taken from a participatory rapid appraisal training manual based on experiences in the Middle East and North Africa [9]

	Households			
	Obi	John	Clara	Unoka
Household members				
Men (number in household)	3	1	0	1
Women	4	2	1	4
Children	5	6	4	10
Labor migrants	2	1	0	3
Animals owned				
Cattle	5	0	0	16
Sheep	24	8	1	56
Goats	15	7	3	16
Chickens	18	23	4	17
Sources of income				
Agriculture	25%	23%	66%	25%
Livestock	17%	8%	17%	21%
Trade and crafts	41%	54%	17%	32%
Remittances	17%	15%	0%	21%
Consumption				
Monthly cash expenses (Dinars)	380	265	85	650

With the data table in hand, such as the one shown in Table 6.1, participants may draw bar graphs and pie charts, such as those shown in Figure 6.2, to compare their ownership of livestock or land, describe the composition of their households or indicate the time they dedicate to various productive activities and the relative income derived from each. If secondary sources of data are lacking, participants can create diagrams on-the-spot to describe their own economic situation, recalling the information from memory. For example, they can draw a circle and divide it into slices which represent the amount of time they spend harvesting forest products, tending to animals, harvesting timber or working in agricultural fields.

A comparison of the results from wealth ranking and livelihood analysis provides you with an overview of the economic status of those community members who gather forest products and how much time the community as a whole dedicates to this productive activity.

6.4.3 Household surveys

After you have formed an initial impression of the local economy from these rapid appraisal techniques, you may carry out a household survey that yields detailed quantitative data on the productive activities of each family in the community. Before beginning the survey, review the techniques that are presented in Chapter 4.

185

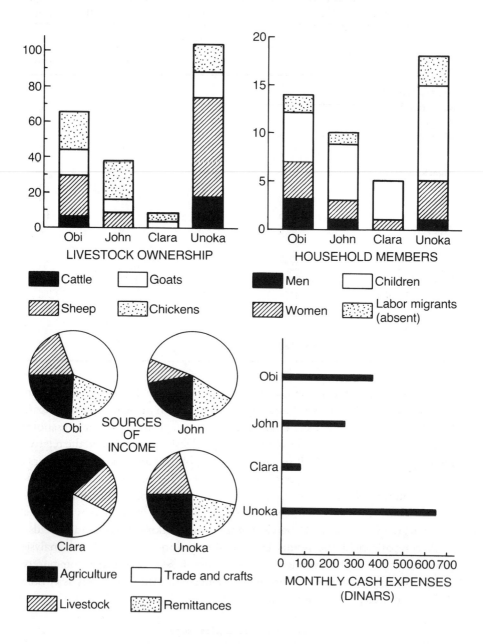

Figure 6.2 Examples of livelihood analysis diagrams taken from a participatory rapid appraisal training manual based on experiences in the Middle East and North Africa. The livelihoods of four families are described by bar graphs which depict household composition, livestock ownership and monthly cash expenses and by pie charts which show sources of income.

Whenever possible, visit all the households in the community or at least a sample which represents local social and cultural diversity. A village census, if available, may offer many basic data on the number and size of the households and land holdings. With semi-structured interviews or written questionnaires, you may elicit details on the amount of forest products consumed and commercialized by each family, the amount of time invested in gathering versus farming or other activities, and total household income, among other data. All data can be compared to the results obtained from PRA techniques to test the fidelity of the various approaches.

A group of researchers from India, headed by K.C. Malhotra, carried out a household survey as part of their study of the role of non-timber forest products in village economies in Midnapore District, West Bengal. As a follow-up to the participatory rural appraisal of sal forest regeneration, described in Chapter 5, they chose to work with 12 Forest Protection Committees, village-based groups that supervise the use of forest resources [103]. After collecting data on the forest and the participating villages, they carried out a semi-structured interview with the heads of 216 households that were selected to represent the socioeconomic and ethnic diversity of the community.

They included the following topics, which can serve as a guide to similar household surveys carried out in other regions: (1) family size; (2) age and gender structure of the family; (3) literacy levels for each member of the family; (4) domestic animals; (5) land holdings and productivity; (6) primary and secondary occupations; (7) annual family earnings; (8) plant species and their parts used by the households; (9) purposes and end uses of the non-timber forest products; (10) seasons of availability and use of each item; (11) market value/opportunity costs of the non-timber forest products; (12) annual consumption of fuel for cooking, parboiling rice, distilling liquor, washing clothes and so on; (13) sources of fuel materials; (14) amount of fodder gathered from the forest; (15) amount of processable material harvested from forests for making minor household articles and (16) quantities of food items gathered from the forest. Some of the practical innovative methods they used to quantify the use of non-timber forest products are described in Box 6.3.

Box 6.3 Methods used in quantifying flow of non-timber forest products at the household level

As part of a study in the Midnapore District of West Bengal, K.C. Malhotra and his colleagues wished to quantify the non-timber forest products gathered by a sample of 216 households belonging to numerous Forest Protection Committees. In a working paper issued by the project participants, they described some methods of estimating the **biomass** (the total weight or volume) and other aspects of the forest harvest:

An emphasis was given on the end use of all forest products and diligent inquiries were made for each household regarding (a) the possible usages of a species, (b) parts used, (c) seasons of availability according to the local calendar, (d) seasons of use, (e) methods of processing, if any, (f) specific purposes of use, (g) whether any customary protection given to the species by the community, (h) possible source of the item, in case not found under their purview of protection, and (i) demand for the item in the local market. This helped generate reliable data on quantities of non-timber forest products.

Biomass of non-timber forest products that are gathered regularly and/or in amounts large enough to meet the subsistence needs of the villagers were quantified. The primary attempts at ascertaining the quantities of items gathered over a season were made by recall method. Approximate amounts were obtained from the ranges of numbers and volumes (basketful, or handful, etc.) given by the informants.

In some cases, however, the recall method could not generate sufficient information. The villagers had obvious difficulties in recalling the quantities they gathered which varied widely with the availability and subsistence requirements of the items. The different species of mushrooms were collected during and after the monsoon in varying amounts: 'sometimes a basketful, sometimes a handful' was the typical answer to queries on quantities. Informants were then supplied with a few clues which proved helpful. For instance, the sale of an item in the market enabled the informant to give a good estimate of the collection. Not all villagers used to sell the item, however. For those who collected the item for household consumption, clues to the range of variation in quantities required for say, cooking a dish for the family, and modal number of days in a season for its consumption yielded sufficiently enlightened guesses. Standard weight of a given volume of a collection was also estimated. Throughout the procedure, suggestive questions were avoided.

The number of animals hunted by the tribals was always uncertain owing to the rarity of the fauna in the jungles. Similarly, the plants occasionally used for medicinal purposes were not quantified because those items are collected in very small amounts as and when required for treating a specific ailment. Quantities of leaves, flowers, etc. gathered from forests for ritual or ornamental use were also not quantified. Wild berries and plums sometimes consumed by the children exploring into the forests were very difficult to quantify and were therefore left out of our estimates.

It goes without saying that a direct measurement of all the non-timber forest product biomass gathered by each sampled household in a season would have yielded more accurate information about the quantities of non-timber forest product flow into the villages. However, this ideal method entails a rigorous year-long survey conducted by a large number of trained fieldworkers in all the sampled Forest Protection Committee villages, and therefore was deemed impracticable, however desirable for our purposes.

The quantifications that we made were in principle based on direct estimation supported by intuitive judgements and cross-checking with as many informants as possible. To standardize the estimates about biomass, several samples of headloads of non-timber forest products were weighed on a spring balance. This method improved the 'guess' about the average weight of a bundle, sackful or basketful of different items of forest produce.

6.4.4 *Calculating return on labor*

Official records, participatory rural appraisal techniques and an economic survey of households will allow you to characterize the various sources of community income, including the amount of days dedicated and money earned per activity. You can make an estimate of the relative contributions to the village economy from products which are traded and the resources which are consumed by villagers.

Another calculation you can make from this data is the return on labor for forest product collection, that is, how much a person can earn per day by gathering a certain plant or animal. As mentioned above, this is a better guide to understanding people's economic behavior than is return from land.

Box 6.4 A forest-based village economy in Gujarat, India

Shashi Kant and Neel Gaurav Mehta, two foresters from India, set out to calculate the contribution of forest products to the economy of Motisingloti, a village of the Bharuch district of Gujarat [101]. Motisingloti is surrounded by moist teak forests that contain timber and fruit trees as well as other plant resources such as bamboo and medicinal plants. The 400 inhabitants of the village, almost all belonging to the Vasva tribe, cultivate a total of 83 ha of agricultural land and collect plant and animal resources from 835 ha of forests. In addition, many villagers seek daily paid labor with the governmental Forest and Irrigation Departments or with local farmers to supplement their income. Combining data from official records and household surveys, the researchers analyzed the portion of village income, the seasonal variations in employment and the return on labor associated with each of these productive activities.

They discovered that forest gathering, which is carried out by 55 of 60 households, occupies a central position in the village economy. Seven non-timber forest products – including seeds, flowers and leaves of various species – are sold by Motisingloti residents to the Gujarat State Forestry Development Corporation. The official price and quantity purchased of each of these products are given in Corporation records, allowing Kant and Mehta to calculate the total income earned from this commerce. Many other forest plants are used for subsistence, including materials for house construction, fuelwood, fodder and wild vegetables. To estimate the monetary value of these products, the researchers multiplied the current market price by the total quantity of each resource that is consumed in the village, as reflected in the household surveys. In many cases, the current market price could be obtained directly from the Corporation or by interviewing vendors in local marketplaces. An estimated price was assigned to non-marketed goods such as edible wild tubers and bulbs.

Based on these estimates, 17% of village income is earned from the sale of non-timber forest products, whereas the monetary value of forest products which are directly consumed by villagers corresponds to 47% of annual earnings. Because an additional 13% is derived from daily labor with the Forest Department, the authors concluded that 77% of the village income is linked to the forest. The remaining 23% is earned from work not carried out in the forest, including subsistence and commercial agriculture as well as daily paid labor in other farmers' fields and with the Irrigation Department.

Through the household surveys, Kant and Mehta also discovered how many days per month each family dedicates to activities which generate cash income. As can be observed in the village totals presented in Table 6.2, 11 390 days are spent collecting forest products and 4425 days working with the Forest Department. This means that out of a total of 17 090 days on an annual basis, 92.5% of the people's time is spent on work related to the forest. This work is clustered in the months of December to July, when most of the yearly cash income is earned. From August through November, villagers work on other farmers' fields or cultivate their own crops.

How well is each type of work paid on a daily basis? The researchers calculated the return on labor (ROL) for forest product gathering by dividing the total income derived by the number of days dedicated to this activity. Although the overall ROL for collection of non-timber forest products is 13.8 rupees per day, this rate varies according to the resource harvested. Bhindi seed, for example yields only 7.2 rupees per day whereas timru leaves bring in 17.6 rupees per day. By contrast, the Forest Department pays 22 rupees per day, the Irrigation Department 14 rupees and large-scale farmers 5 rupees. These figures help explain the monthly distribution of labor shown in Table 6.2. Only when there is no possibility of working with the Forestry and Irrigation Departments or of collecting forest products do villagers accept the lower wages that come from working in other farmers' fields.

Table 6.2 Monthly distribution of work days for different income earning activities in Motisingloti, India

| Month | Forest Dept. | Irriga-tion Dept. | Other farmers' fields | Non-timber forest products collection | | | | | Total | |
				Mahua flower	Mahua seed	Timru leaves	Sag seed	Bhindi seed	Forestry activities	Grand
January	-	-	-	-	-	-	262	266	528	528
February	-	-	-	126	-	-	104	1064	1294	1294
March	1100	-	-	506	286	-	-	1064	2956	2956
April	1100	-	-	885	1145	564	-	266	3960	3960
May	1100	-	-	632	1145	1972	-	-	4849	4849
June	500	350	-	253	286	282	-	-	1321	1671
July	625	400	100	126	-	-	-	-	751	1251
August	-	-	100	-	-	-	-	-	-	100
September	-	-	50	-	-	-	-	-	-	50
October	-	-	50	-	-	-	-	-	-	50
November	-	-	110	-	-	-	-	-	-	110
December	-	-	115	-	-	-	156	-	156	271
All	4425	750	525	2528	2862	2818	522	2660	15815	17090

Let's take an example from a study carried out in Gujarat, India, described in Box 6.4, to see how this works. From their household surveys, Shashi Kant and Neel Gaurav Mehta discovered that the average villager from Motisingloti spends a total of 60 days per year, working 5 hours per day, to collect a total of 960 kg of Bhindi seed. With a single workday defined as a period of 8 hours, they concluded that a total of 37.5 workdays are dedicated to Bhindi seed gathering (300 hours ÷ 8 hours per day = 37.5 workdays) plus an additional 3 days per year for processing, giving a total of 40.5 workdays. The seed must be dried before selling it to the Corporation, reducing it to 10% of its fresh weight. The dry seed is purchased at the rate of 3 rupees/kg yielding an income of 288 rupees (96 kg of dry seed × 3 rupees per kilo = 288 rupees). Dividing by the 40.5 days invested in collecting and preparing this resource, we can see that this activity yields a little more than 7 rupees per workday (288 rupees ÷ 40.5 workdays = 7.1 rupees per workday).

6.5 Local markets

6.5.1 Surveys of biological resources sold in local marketplaces

No economic study of biological resources can be complete without a detailed survey of the plants and animals sold in local marketplaces. Many medicinal plants, ornamentals, wild foods, edible insects and other products have a strictly regional value which can only be discovered by talking with producers, sellers and consumers. The amount of income that a community can earn from trade in these products is determined by their access to marketplaces and the current value of the goods as determined by supply and demand.

Carol Smith, an American anthropologist who carried out a detailed study of rural marketing systems in western Guatemala, suggests concentrating on four elements of marketplaces. Begin by documenting the diversity of **commodities**, which in an ethnobiological study would correspond to the plants, animals, mushrooms, insects and other biological resources which are offered for sale.

Note the different types of **vendors**. In Guatemala, for example, part-time vendors come to the marketplace not only to sell produce harvested in their own community but also to buy goods from other regions. They interact with many full-time vendors, including local middlemen who import items into the region as well as long-distance traders who come to sell goods they may bring from far distances. In many parts of the world, the majority of merchants you encounter will be women but men also play an important role in marketing some goods. The vendors usually represent a cross-section of different social classes and ethnic groups.

Pay attention to the types of **economic transactions** which take place, including barter, the exchange of one kind of good for another, and selling for cash. There will be wholesale exchanges of bulk produce destined for redistribution outside the region where it was harvested. You will also find many retailers who sell small quantities of both imported and local goods. Some products which are produced in rural areas move upward through a chain of marketplaces until they reach their final destination in urban areas. Other merchandise is purchased in the city and eventually filters down into smaller and more remote villages through marketplace transactions.

Finally, document the **general geographical setting**, including the ecological zones from which the marketed goods originate and the number, size and variety of communities which rely on the market as an outlet for their produce as well as a source of imported goods. Calculate how far these communities are from the marketplace both in terms of actual distance and the amount of time required for travel. You will find that the market serves as a meeting place for a broad range of people from many communities and as a point of redistribution for the products of many different ecological and cultural zones.

6.5.2 Observing marketplace interactions

When making a market survey, get an early start and plan to stay throughout the day. Many forest products brought in small quantities are likely to disappear soon after the first buyers begin shopping. Fresh plants sold at the beginning of the day make the best collections and vendors are more apt to engage in conversation before the market reaches its peak activity. A bustling marketplace at midday provides other opportunities, including observing the great diversity of consumers and vendors as well as displays of crafts and other non-perishable goods of ethnobotanical interest.

When you arrive, take a quick walk through the entire marketplace to get a feeling for its layout and to spot the goods you wish to purchase. Make a rough sketch of where different types of vendors are located and take time to observe the kinds of transactions taking place between vendors and purchasers.

If you listen to conversations in a marketplace, you will find that buyers pose the same questions as any curious ethnobotanist. When they spot a fruit they have not seen before, they ask what it is called, where it comes from and how to eat it. If they are purchasing a medicinal herb, they want to know which illnesses it will cure, how to prepare it and for how long to take it. You can learn much by paying attention to the items which are purchased and the prices paid. When appropriate, you can ask them some of the same questions you pose to merchants, as described below.

After observing the range of products offered for sale and how others shop in the market, continue your market survey by becoming a buyer yourself. Wander from stall to stall, engage vendors in conversation, ask to taste fruits, inquire about the names and uses of plants, make purchases and bargain for lower prices. You will probably soon find yourself going to the market to purchase food, cure your ills and adorn your house just as other consumers do. This integration into local patterns and rhythms of shopping will give you special insight into how the market functions.

The most critical aspect of a marketplace survey is establishing rapport with vendors. Because they spend the entire day selling produce, they are accustomed to answering consumers' questions. If the vendor asks why you are so curious about plants, take the opportunity to tell him or her briefly about the study you are carrying out. When you begin to interview certain vendors, buyers or producers in depth and to visit their communities, you should explain in detail the aims of your research to them and to local authorities.

6.5.3 Recording ethnobotanical data from marketplaces

Unlike a typical buyer, you should be systematic in your enquiries, asking for the same information about each purchase. Before venturing into the market, decide on the questions that you will ask. The data you collect on the goods, vendors and other aspects of the marketplace can be jotted down in a notebook throughout the day. These notes can be transferred to a pre-printed form, such as the one shown in Figure 6.3, while you are preparing the collections at the end of the market day.

In many respects, conducting a survey of useful plants in a marketplace is similar to carrying out an ethnobotanical inventory in a community. Some whole plants are sold and may be pressed and dried as herbarium specimens. Other vouchers, including most fresh fruits and vegetables, should be preserved in a spirit collection. Many specimens need little preparation, including dried bundles of medicinal

Projek Ethnobotani Kinabalu **Marketplace Collections**

Location of Marketplace: _____

Collector: _____ Collection Number: _____ Date: / /

INFORMATION ON THE VENDOR

Name: _____

Type of Vendor: Permanent stall   Temporary stall   Ambulatory

Village of Vendor: _____ Gender: male female Estimated age: _____

How often do they sell here? _____   In other markets? _____

INFORMATION ON THE COLLECTION:

Local name: _____ Lifeform: _____

Village: _____ Vegetation type: _____

Cultivation status: cultivated   managed   wild

Marketing status: gathered by vendor   resold

Number of species in collection: single   mixture of _____ plants

Names of other ingredients: _____ _____ _____

Condition of plants: fresh   dried   preserved in _____

Price/unit: _____ Brought to market: daily   weekly   on occasion

Estimated quantity: vendor _____   whole market _____

Availability: jan feb mar apr may jun jul aug sep oct nov dec

How much sold now compared to in past: more   same   less

Why?   less available for harvest   less demand by buyers   Other _____

Use: _____

Plant part used: _____

Preparation: _____

Notes: _____

HERBARIUM INFORMATION:

Botanical family _____ Scientific Name: _____

Preparation: herbarium specimen   spirit collecton   ziploc bag

Distribution: Kinabalu Park (KP)   Sandakan (SAN)   Kuala Lumpur

(KEP)   Kew (K)   Leiden (L)   Other: _____   Total number of

duplicates: _____

Figure 6.3 A pre-printed form designed to record information on ethnobotanical collections from marketplaces around Kinabalu Park.

herbs, bags of spices and packages of seeds, all of which can be stored in plastic bags. Preparation, labelling and distribution of the collections can proceed as described in Chapter 2. Objects which are handcrafted from plant or animal materials, including baskets, textiles and wood carvings, may require specialized documentation [104].

You will probably wish to photograph all collections before preparing them, taking special care to document the form in which they are sold – dried or fresh, bundled or loose, separate or mixed with other species. These color slides or black and white photographs complement shots you take in the marketplace itself, depicting the display of the produce, the merchants and consumers, the architecture of the stalls and other relevant features.

Some special measures need to be taken when preserving market specimens, such as treating all collections immediately for insect infestation. Although not always visible, many small insects and eggs may be found in dried fruits and seeds purchased in the market. These will be killed automatically if the collection is dried over fairly high heat or pickled in a preservative such as alcohol. Herbs, seeds and other specimens which are already dry may be heated for 24 hours in a plant dryer.

If you are collecting germplasm or other material that must be kept viable, you will have to use an insecticide or other method of treatment that does not kill the plant material you plan to propagate. Some researchers collect extra live material, such as seeds, roots or stem cuttings, which they grow in a garden or greenhouse. If the specimen eventually flowers and bears fruit, a fertile voucher specimen can be made, allowing botanists to make a definitive determination of the collection.

How many specimens do you need to collect in order to ensure a complete inventory? The answer will depend on the ecological diversity of the area in which you are working and the number of people serviced by the marketplaces you visit. In an ethnobotanical survey of marketplaces of the Central Valleys and Sierra Norte of Oaxaca, several local university students and I collected 651 voucher specimens, including 626 vascular plants (606 flowering plants, eight conifers, two cycads, ten ferns), 17 mushrooms, seven bryophytes and one lichen [105]. These collections, which include some 97 families, 256 genera and over 300 species of plants, probably constitute a relatively complete inventory of locally produced plants which are sold in the Valley and Sierra markets, systems which draw goods from many different ecological zones and serve the needs of hundreds of thousands of people who live in rural communities and urban areas.

When you study the local marketing of plants and animals, plan on visiting the marketplaces periodically during at least a one-year period. This will allow you to record the seasonal variation in what is offered for sale. Although many spices, medicinal herbs and staple foods are found year-round, other plants come and go with the ebb and flow of supply and demand.

6.5.4 Quantifying supply, demand and value of marketed products

Besides making opportunistic purchases during periodic visits to marketplaces, there are other approaches to making a systematic appraisal of market ethnobotany. These techniques are time-consuming but offer a unique perspective on the quantity and diversity of goods that flow through a marketplace. If you have become acquainted with some vendors particularly well, they may allow you to make a complete inventory of their stalls. You can count how much they have of each product, record all the prices and arrive at an estimate of the total amount of each plant that is sold on an annual basis and the total value of their complete stock of goods. Robert Bye and Edelmira Linares, who have spent many years studying plants sold in Mexican markets [106], were able to conduct complete surveys of several medicinal plant stalls in the *Mercado de Sonora* in Mexico City. This systematic approach allowed them to document the flow of goods through the market, which complemented data they collected as they were making random purchases.

Another way to quantify your data is to count the number of vendors who are selling a particular species. In addition, estimate how much is offered for sale in each stall and record the price per unit. This will allow you to calculate the total supply that is available on a particular market day as well as the overall value of the resource. If you make one assessment as the day begins and another as buyers are packing their wares to return home, you can obtain a rough measure of the total demand, that is, how much was purchased by consumers. A series of these measurements over the course of the year will allow you to calculate fluctuations in supply, demand and price as well as the total amount and value of the resource on an annual basis. Even if you are not able to obtain a quantitative estimate of supply and demand, you can note the relative availability of each species. For example, Robert Bye and Edelmira Linares used a six point scale, ranging from 0 (absent) to 5 (abundant), when gauging the seasonal variation in availability of produce in the Mercado de Sonora.

6.5.5 Visits to producer communities

After documenting plant and animal resources in the marketplace, you may wish to visit the communities in which they are harvested. Accompanied by local people who gather the resource, you can visit the exact site where the harvesting takes place, allowing you to collect voucher specimens of the various species which are marketed. They can help you to determine the quantity of material which is taken from the forest each year and the number of people involved in the harvest. They may also be able to give you information on neighboring communities which participate in the trade.

A rapid appraisal of the village economy will allow you to understand the role played by extractive activities. Communities and households that earn sufficient income from cattle-raising, fishing, plantation agriculture or temporarily migrating to sell their labor typically do not need to engage in marketing of medicinal herbs, handicrafts or other minor goods. As part of this appraisal, calculate the distance from the community to the marketplace and try to obtain an estimate of how much it costs to transport produce by the different means available. Ask if there are middlemen who purchase some of the production in the community before it reaches the marketplace and what price they pay for various resources. Determine if the village specializes in production of goods which are not generally available in surrounding communities, such as fruits available from plants of restricted distribution or handicrafts that are unique to a certain cultural group. Because participation in extractive activities can vary from household to household, attempt to discover the number of families who are involved in forest products trade and their relative status in the community. Participatory rural appraisal techniques such as wealth ranking will allow you to understand these socioeconomic dynamics in a short period of time.

A visit to producer communities provides a good opportunity to assess if the harvesting is sustainable or if, on the contrary, it is threatening the survival of the biological resource and local livelihoods. You may observe special techniques used by people to ensure steady production over many years. Ask local harvesters for their impression of the relative abundance of the resource. Is the yield diminishing or increasing? Is the resource more accessible now or is it necessary to travel further from home to harvest the produce than in the past?

6.5.6 The market survey

Few researchers have the time to visit all the harvest sites of the entire range of plants, animals and insects that are sold in a market. For this reason, it is important to record detailed information on each purchase you make while you are in the marketplace. Collect the same data as for any ethnobotanical voucher specimen, including local name and lifeform, use, plant part employed, mode of preparation and dosage, among other points of information discussed in previous chapters of this manual. When possible, interview both consumers and merchants, paying close attention to any patterns of variation in the names, uses and other attributes of the various species you encounter.

Keep in mind the following additional lines of enquiry, illustrated with examples from the ethnobotanical market survey we carried out in the Central Valleys and Sierra of Oaxaca:

- **Information on the vendor.** Note whether the vendors sell from a permanent stall, a temporary stand or if they are ambulatory – walking through the markets selling their produce. Are they selling goods at a retail or a wholesale level? Do they specialize in medicinal, edible, ornamental or ritual plants or do their goods cross-cut these categories? Record the same sociological information as you would for the participants in a community survey, including their name, age, ethnicity, home town and other personal details. Ask how long they have been selling in the marketplace, if they come regularly and if they sell elsewhere.

  Through our series of casual interviews with vendors in Valley markets of Oaxaca, we were able to confirm that Zapotec Indian and Mestizo vendors control the local commerce in forest and agricultural commodities. Members of Chinantec, Mixe, Mixtec and other indigenous communities play more minor roles as buyers and sellers in the marketplaces.

- **Origin of the produce.** Ask for the name of the village or region where the resource is harvested. Enquire about the ecological and climatic conditions of the source location. Determine whether the species is gathered from wildlands in the community or from cultivated or managed zones. An overall perspective of the hinterlands (the region around the market town) can be obtained with large-scale climate and vegetation maps of the region.

In the case of Oaxaca, we discovered that produce comes from the coast, isthmus, mountains, gulf lowlands and other regions of the state. This covers a diverse range of ecological zones, including tropical evergreen forest, cloud and pine-oak forests and tropical deciduous woodlands. As in many parts of the world, marketplaces in the Valley and Sierra serve as centers of distribution for products of many diverse areas.

- **Condition of the goods.** Is the produce sold fresh, dried or preserved in another way? How is it packaged? Is it sold alone or combined with other ingredients? When collecting dried or fresh whole plants, note whether the specimens are sterile or fertile.

  Of the 651 voucher specimens that we made, 260 were whole plants that we pressed as herbarium specimens. The other 391 collections consisted of dry material that we stored in plastic bags or fresh plant parts that we pickled in alcohol. The majority of plants that we encountered were sold separately but mixtures of medicinal herbs were common.

- **Management and marketing of the resource.** Ask if the resource is cultivated, managed or wild. Do the vendors gather the produce themselves or do they purchase it for resale from someone else?

  Although many of the cultivated plants sold in the Oaxaca Valley markets have been passed from producers to middlemen to vendors, managed and wild plants are often brought by mountain dwellers who set up temporary stalls on the edge of the market. They and their family members gather most of these species – including ornamental cycads, flowering orchids and medicinal herbs – in forests and fields of their community.

- **Quantity, price and availability.** When possible, count the number of vendors that carry the product and estimate how much is being sold in each stall. Note the price per unit and how this varies among different vendors. Ask if the produce is available year-round, if it grows scarce during some weeks or is found during only a short period of time. Is it brought to market daily or only on occasion? During subsequent visits to the market, attempt to verify these seasonal and temporal patterns.

  In the Central Valleys of Oaxaca, the immature inflorescences of the tepejilote palm (*Chamaedorea tepejilote* Liebm.) are available for only a few months in early spring, when the flowering shoots are emerging. Bromeliads are available in mountain forests all year but are sold as ornamentals in the markets only during festive periods such as Christmas and Easter, which also corresponds to the time when they flower. In contrast, dried medicinal herbs are in constant supply throughout the year.

- **Changes in demand and supply.** Ask the vendors if the supply of a particular species is dwindling or increasing and if they can suggest why this is happening. If they harvest the resource themselves, enquire how long they have to travel from their home to find the plants and animals and if the distance or time is increasing. Note whether consumers purchase more or less than in previous years.

On a market day in Ocotlán, we talked with ambulatory vendors from Santa Cecilia Jalieza, a Zapotec village in the Central Valleys of Oaxaca, who are frequent visitors to local marketplaces. They search out market-goers, particularly tourists, to whom they sell handcarved items such as letter-openers. These are made from a wood called *yagalán*, which comes from *Wimmeria persicifolia* Radlk., a tree in the Celastraceae. Demand for the objects appears to be growing as the number of tourists increases. The *yagalán* trees of Santa Cecilia disappeared long ago, so the artisans now buy their supply of wood in the marketplace of Ocotlán. The suppliers of the wood, who come from a distant community called San Miguel Peras, have to walk increasingly further in the forest to find *yagalán* trees. The wood-carvers of Jalieza are now using some substitute lower-quality woods.

- **Additional information.** In addition to these questions, leave a space on the information sheet where you can record the collector's name, the collection number that corresponds to the specimen, the name of the market community and the date of purchase of the goods. You may also wish to leave room to record notes on the preparation, distribution and scientific identification of the specimens. For botanical collections, note whether the specimens have been pressed as herbarium specimens, preserved in spirit or placed in sealed plastic bags. Note the total number of duplicates and in which herbaria or botanical museums they have been deposited. Once the specimens have been identified, record the botanical family and complete scientific name, including varieties or subspecies when appropriate.

This information can then be transferred to a computerized database to facilitate analysis of the collections. In addition to the categories drawn from the information sheet, fields can be added to record the elevation, climate, coordinates and ethnicity of producer communities. You can also add data about the species that have been collected, including their overall distribution and whether they are native or exotic species.

The database on ethnobotanical collections of Oaxaca marketplaces reveals that 32% of the plants are primarily used as medicinals, 31% as food, 23% as ornamentals and 8% as condiments. The remaining 6% are sold for diverse other purposes including household utensils and plants used ritually. Sixty five per cent are native species and 35% exotic. We found the largest percentage of native species among the medicinal plants (76.1%), followed by food plants (59.6%), ornamentals (55.7%) and condiments (47.9%).

The information obtained from visits to marketplaces and producer communities can be used to assess the value and conservation status of the species detected in the marketplace survey, as exemplified by the case study presented in Box 6.5.

Box 6.5 Documenting the value and conservation status of species detected in market surveys

Jerzy Rzedowski, a Mexican botanist, discovered a new species of piñon pine (*Pinus maximartinezii* Rzedowski) through a market collection that he made in the village of Juchipila, Zacatecas in the 1960s. By speaking with local vendors, he was able to discover the only known population of this pine on a small mesa in the Sierra de Morones. The trees belong to three land owners, who make a yearly harvest of the pine nuts and market them in the region.

Zsolt Debreczy, a Hungarian forester who visited the site in 1992, interviewed Antonio Benavides Gómez, one of these men. Owner of 40 hectares of *Pinus maximartinezii* forest, he harvests 50 large bags of pine nuts in the month of September with the help of family members. Each bag weighs around 25 kg and sells for the equivalent of about US$125 at 1992 prices. This activity brings in over US$6000 for the Benavides family, nearly four times the annual income of most local people.

Although it appears that the current intensity of harvesting has been sustained over many decades, there is evidence that this and other resource use patterns may be having an adverse impact on the natural regeneration of the species. Jesse P. Perry, Jr, a specialist on the pines of Mexico and Central America, notes that during a visit in the 1980s, he found no evidence of reproduction of this piñon pine. Fire, grazing by goats, burros and cattle and perhaps over-harvesting of the pine nut trees are the main threats to the survival of this relic population.

7

Linguistics

Figure 7.1 Rajibi Haji Aman, deputy director of Sabah Parks, teaching participants in the Projek Ethnobotani Kinabalu how to transcribe Dusun, the language spoken in most communities around Kinabalu Park. Rajibi and other members of the park staff who are native Dusun speakers are participating in an inventory of local plant names and uses.

7.1 Learning a local language

When you begin working in a new community, you may have the opportunity to learn the language spoken by the local people. Of the world's estimated 6000 living languages, most are spoken by small ethnic minorities ranging from a few individuals to several thousand people. These languages, many of which are not traditionally written, are often spoken in the communities where ethnobotanists tend to work. Few courses are offered in these languages and even simple vocabularies and grammars may be hard to obtain.

If you plan to live one or more years in the region, you may wish to spend some time every day studying to speak and write the local dialect. Knowing a language adds much depth to ethnobotanical research. As local people realize that you are beginning to understand their culture and dialect, they are apt to discuss their knowledge of the natural environment in greater detail. Even if you are a native speaker and a member of the community, you may have to learn to write your language in a way that linguists and others can understand.

Learning an unwritten language requires special skills that you should try to acquire before going into the field [107]. Consider taking an introductory course in linguistics or a specialized seminar on how to write and pronounce the various sounds that are found in the languages of the world [108]. Keep in mind that language learning involves three basic processes: (1) understanding what people say, (2) expressing yourself in conversation and (3) writing correctly. You may find that your progress in one area is more advanced than in another. For example, you may be able to write and re-read specific words or phrases before you are able to grasp the meaning of speech or participate in a conversation. Your comprehension can be aided by studying any grammar books or dictionaries that you are able to obtain, but the bulk of your effort will focus on patiently listening, repeating and writing.

Many fieldworkers choose a single native speaker to guide their language learning. The instructor is often bilingual, which allows the outsider to obtain some explanations and translations of native terms in a language he or she can easily understand. While it is productive to maintain a steady interaction with one or more instructors, you should attempt to speak with other native speakers in order to gauge your progress and grasp variation in the way the language is used.

7.2 Collaborating with linguists

Many ethnobotanists feel they do not have sufficient time or adequate skills to learn a language while carrying out fieldwork. Just as they rely on pharmacognosists to analyze the chemical properties of plants and on botanists to identify their collections, they seek the assistance of linguists in order to write and interpret key words and concepts encountered in the course of their research.

As you begin to search for a linguist who can accompany you in the field, you will likely find several types of potential collaborators. You may discover that there are university linguists who are specialized in one or more languages of your study area. But if you visit a university linguistics department, do not be surprised to find many researchers dedicated to contemporary approaches such as transformational linguistics or sociolinguistics, topics which are on the cutting edge of linguistic theory. Few university professors have the opportunity to live many years with local people, a necessary condition for pursuing the descriptive and historical approaches which are of most interest to ethnobotanists.

Missionaries, who often spend many years in a single community, are in a privileged position to study local languages in great detail. There are a large number of religious workers, affiliated to the Summer Institute of Linguistics and other international Protestant organizations, who have received basic or advanced linguistic training. One of their main goals is to translate the Bible into the various dialects of a language. To accomplish this, they develop practical writing systems, publish extensive vocabularies and often produce detailed analyses of grammar, phonology and semantics of many previously unstudied languages. Because they often combine their linguistic studies with efforts to convert local people to a Protestant religion, the presence of these missionaries in many countries is controversial to local people and governmental officials. Yet in other areas, their efforts to teach people to read and write local languages have led to general acceptance of their work. Because it is a sensitive issue, it is appropriate to ask local people and other co-workers if they agree to collaborate with missionary linguists.

You may discover that some local people have a special facility for teaching you their language. They may already be experienced in transcription, a skill acquired either through their own efforts or through training received from governmental literacy programs, missionaries or academic linguists. Local people often possess an intuition for analyzing the meaning and pronunciation of words that is superior even to that of non-native linguists who have spent many years in the community. Local people are in a better position to continue studies well after outsiders have left the study area, refining linguistic analyses and recording new names for plants and animals. Keep in mind that unless these local assistants have received extensive training in linguistics, their methods of transcription and analysis may be less rigorous than those of academic linguists and missionaries.

Even if you seek the collaboration of someone who has linguistic training, you should try to gain practice in hearing, pronouncing and writing the distinctive sounds of the language in which you are working. This exercise will give you a better grasp of the way linguists are writing the language and will allow you to tune your ear to hear the distinctive speech sounds of the language.

7.3 Where there is no linguist

You will not always have the luxury of working in the field with a trained linguist. If you are unable to learn the language yourself, how can you ensure that linguistic data is gathered and transcribed in an accurate way? The best solution is to carry a tape recorder to the field and to make recordings that you can later pass along to linguists specialized in the local language or that you can analyze yourself as you gain linguistic expertise.

As you identify the key words that you wish to transcribe, you can ask one or more native speakers to pronounce each term separately. It is best to have the person pronounce the word once slowly and then again in a normal tone and speed of pronunciation. You may also ask him or her to pronounce the word in a phrase, such as 'We use x as a remedy for stomachache' or 'x is a type of tree'. This will allow the collaborating linguist to hear the distinct sounds clearly and to understand how the word is pronounced in everyday speech.

Even if you are relatively proficient in transcribing the local language, you should record plant and animal names that you discover during the course of your research. The tapes act as a linguistic voucher, much in the same way that plant specimens document botanical species. By leaving the tapes in a museum or other research center, you will ensure that colleagues can refer to your work, verifying the accuracy of your transcription.

An efficient way to tape record plant names is to work with a local person and a set of ethnobotanical voucher collections that have been collected in the community, either by you or a collaborator. Pronounce the name of the collector and the number of each specimen and then ask a native speaker to say the name of the plant. As noted above, the first pronunciation can be carefully enunciated to allow recognition of the component sounds, but additional pronunciations should be articulated normally as if the name were being used in everyday speech. Because each name is linked to a voucher specimen that will be identified by botanists, this system allows you to establish the correspondence between folk and scientific categories in an efficient way.

Once you have the tape recorder in hand, you may wish to transcribe additional information about each voucher specimen, such as the lifeform, uses and preparation as well as any myths or stories which refer to the plant. Avoid the temptation to depend too much on recording information that you expect to transcribe and analyze once you leave the field. Keep in mind that it will take many hours to transcribe a 60-minute tape and much additional time to interpret and translate the recording.

7.4 Transcribing the local language

The process of recording languages such as Mixe and Chinantec using written linguistic symbols is referred to as **transcription**. In order to record speech accurately, phoneticians analyze the pronunciation of each distinctive sound unit, referred to as a **phoneme**. They also seek to establish a set of rules which explain how these sound segments change when they occur in relationship with other phonemes in words and phrases. **Phonology** is the study of the sounds and the set of rules that determine how words and sentences are pronounced.

In order to transcribe a language in a way that can be understood by other colleagues, phoneticians use the standardized symbols of the **International Phonetic Association** (often abbreviated as IPA). Linguists who focus on a particular region of the world sometimes employ slight modifications of this system which are widely understood by their other researchers in the area. Linguists who are developing a writing system (a local **orthography** that can be used by non-linguists and native speakers) may use an even simpler set of symbols that are more easily understood than standard linguistic notation.

In many popular books on ethnobotany, the authors include a key to pronunciation which list the orthographic symbols (the letters used to write the language) in alphabetical order. Richard Felger and Mary Beck Moser, in their book on the ethnobotany of the Seri Indians, show the Seri orthographic symbols that they use and give the equivalent International Phonetic Association symbols [109]. They also provide examples in English which show how to pronounce most of the Seri symbols.

7.5 Linguistic analysis in ethnobotany

7.5.1 Analyzing the meaning of words

While learning to correctly transcribe the local language, you should also seek to understand the meaning of the plant names and other terms used to describe the natural environment [110, 111]. It is surprisingly common to find articles on ethnobotany in which the authors make no attempt to give even a rough translation of the local name. There are even stories of researchers who have recorded apparently local names of plants, only to have other ethnobotanists discover later that the local people were simply saying 'I don't know' or 'That's a rock'. These misunderstandings can be avoided by analyzing with local people the meaning of each plant name.

Linguists differentiate between literal and free translations. Free translations, also called 'glosses', are the closest English (or other language) equivalent to a native term. Literal translations are word-for-word renditions. For example, the free translation for the Mixe plant name *tsapxòj* is *quince* (a type of fruit tree), while the literal translation is *heaven-oak*. *tsap* literally means *heaven*, but is commonly used as a modifier for objects introduced by the Spanish. *xoj* is the general name for oaks, which are native. Together, the two terms are used to refer to *Cydonia oblonga* Miller (quince), a fruit tree introduced by the Spanish that apparently reminded some Mixe speakers of native oaks.

In many regions of the world, you can make an initial translation of plant names and other terms by seeking assistance from a native speaker who is fluent in a major non-indigenous language such as French, Spanish or English. You may also consult bilingual dictionaries of the local language that can help you understand some of the names. When you seek to translate local terms into a broadly-spoken language such as English, Spanish or French, keep in mind that the exact equivalent may not exist for many concepts. The Italian saying, '*traduttore, tradittore*' – which means 'translator, traitor' – is meant as a warning that the process of defining in one language the meaning of terms in another is always somewhat imprecise.

You can sometimes gain insight into the meaning of names by breaking the word down into its component parts. For example, if you were to hear **tsapxoj** in a Mixe community, you could understand the meaning of the name by asking 'What does **tsap** mean?, What does **xoj** mean?'. If you have a list of names, look for similar suffixes, prefixes or root words. By asking what the names have in common, you can often find clues for interpreting their meaning. For example, there are many Mixe plant names that begin with **tsap**. By considering them as a set, I realized they were all introduced species that resembled native plants in some way.

At times, you will have to consult technical linguistic works to interpret some terms. In the ethnobotanical inventory of Totontepec, I discovered several plant names, such as **aaydum** (*Annona cherimola* Miller) and **maydum** (*Mimosa albida* Kunth), that end in '**-dum**'. The people of Totontepec could not interpret the suffix, but a systemic analysis of many Mixe dialects carried out by a Danish linguist revealed that '**-dum**' is an old Mixe suffix which refers to round things – an appropriate descriptor for the spherical fruits of *Annona* and the ball-like inflorescences of *Mimosa albida*.

Keep an eye out for plant names which have been borrowed from other languages. While analyzing Dusun plant names with native speakers in northern Borneo, for example, I discovered that **bayabas**, the Dusun name for guavas (*Psidium guajava* L.), was derived from the Portuguese word *guayabas*. Introduced plants, such as the guavas brought to Malaysia by the Portuguese, often carry borrowed names. Yet even native plants can be labelled by borrowed names in situations where there has been a long history of contact between two cultures.

Avoid the temptation of becoming overly creative in your study of the meaning of names. If your interpretation cannot be confirmed by local people and linguists, do not be convinced that you are right. Keep in mind that the names of many plants and animals are proper names that refer specifically to the biological organism and cannot be analyzed further. Imagine if someone asked you to explain the meaning of 'maple' or 'oak'. These names have no other meaning for English-speaking people in North America or Europe; they are merely labels given to different species of hardwood trees. A large portion of the names given to distinctive, significant species in any culture will be similarly unanalyzable.

Providing an accurate translation of names is only the first step in understanding their meaning. You will also seek to discover the **referent** or object to which each name refers. In the case of plants, for example, you would find out which species or range of species correspond to each local name, a subject discussed in greater detail below.

Keep in mind that this referential meaning is only one aspect of a semantic analysis of folk classification. Meaning is also derived from how the plant is used and perceived in the community, including its symbolic importance as portrayed in myths and rituals. The first steps of accurately translating a name and determining its referent opens a door to gathering detailed ethnographic data that explains its overall meaning and significance in a culture.

7.5.2 Detecting cognates

Terms from different dialects or languages are said to be **cognate** when it can be shown they are variations of the same word and refer to the same object. If you are working in more than one community in a region where people speak the same or related languages, you will find that many plant and animal species are called by a similar name in each locale [112]. Although some of these similarities may have arisen coincidentally or through linguistic borrowing from one community to another, many of these names can be considered true cognates.

Søren Wichmann, a Danish linguist who has studied the relationship between Mixe and Zoque languages of southern Mexico, has encountered many plant names which are cognate throughout the region. *Erythrina*, a genus of trees in the Fabaceae, is called **tsejst** in Totontepec, where the Northern Highland Mixe dialect is spoken. In Southern Highland Mixe and Midland Mixe, the trees are referred to as **tsejch**, whereas speakers of Lowland Mixe use the term **tsejchk**. By observing all these forms together, linguists are able to reconstruct a probable ancestral term from which all the variants have been derived.

This ancestral term forms part of what linguists refer to as the **protolanguage**, the parent language that was spoken – usually thousands of years ago – before it split into the numerous dialects that exist today. Søren Wichmann has reconstructed the proto-Mixe term for *Erythrina* as ***tsejch**, which is the same as the word used in the present day Southern Highland and Midland dialects of Mixe.

Note that reconstructed terms from a protolanguage are always preceded by a superscript asterisk $^{(*)}$, indicating that they are assumed to be the original form of the word, but that no linguist has actually heard them pronounced. Although you will probably not have the opportunity to reconstruct the ancestral forms of plant and animal names while you are in the field, you should ask collaborating linguists if these data are available in any historical studies of the language.

Ethnobiologists are interested in reconstructed terms because they can tell us, along with the results of archaeological studies, which plants and animals were important in a region thousands of years ago. Linguists assume that if a word was part of the common vocabulary of the speakers of a protolanguage, then the object or concept referred to must have been significant to the local culture of that period. As the ancestral population split into numerous isolated communities and the original language evolved into distinct dialects, these key words were maintained.

Because plant and animal populations have characteristic geographical distributions, linguists can postulate the **homeland** or original territory of an ethnic group by studying the reconstructed terms of protolanguages. Human populations have migrated over thousands of kilometers, but the cognate terms in their languages can indicate that they were familiar with the flora and fauna of a specific region – their postulated homelands – thousands of years ago.

Whether or not you decide to delve into the history of the language, detecting cognates can reveal much about the contemporary classification and use of plants in linguistically related communities. If this approach is appropriate for your ethnobotanical study, how can you proceed to identify cognate terms? First prepare a fairly complete inventory of plant names for various communities and determine which species correspond to each folk category. Using either a card file or a computerized database, make a table ordered according to scientific species and which contains the local names from each community.

An example drawn from the Projek Ethnobotani Kinabalu is shown in Table 7.1. We asked Dusun speakers from five different communities to give us a list of all the non-rattan palms they knew. John Dransfield – a specialist in Southeast Asian palms from the Royal Botanic Gardens, Kew – provided a preliminary scientific identification after discussing the morphological characteristics and uses of each folk category with the participants. For some species, such as the betel nut palm (*Areca catechu*), the exact same name is used in all the communities. Other species have names which vary slightly, but are still cognates – *Arenga brevipes* is called variously **karumbohoko** or **karumohoko**. Still other species have distinct names which are not cognates. The various species of *Salacca* in the

Table 7.1 The results of a free listing of non-rattan palm names given by Dusun participants from five communities in the first training course of the Projek Ethnobotani Kinabalu. Parentheses around a Dusun name indicate that the palm does not grow in the community, but residents know the name and are familiar with the plant. Solidus (/) indicates that there is apparently no name for the palm in the community

	Kiau	Bundu Tuhan	Poring	Serinsim	Takutan
Areca catechu	lugus	lugus	lugus	lugus	lugus
Areca kinabaluensis, Pinanga spp.	bumburing	bumburing	bumburing	bumburing	bumburing
Arenga brevipes	karumbohoko	/	karumohoko	/	karumohoko
Arenga undulatifolia	(polod)	(polod)	polod	polod	polod
Caryota mitis	betu	betu	betu	/	betu
Caryota no	/	/	/	giman	giman
Cocos nucifera	piasau	piasau	piasau	piasau	piasau
Eugeissona utilis	(luba')	(luba')	luba'	luba'	(luba')
Licuala spp.	/	(silad)	silad	/	silad
Metroxylon saga	rumbioh	rumbioh	rumbioh	rumbioh	rumbiyoh
Oncosperma horridum	(telibung)	(telibung)	melugus	melugus	telibung
Salacca spp.	terintid	terintid	begung	begung	begung
unidentified	toyon	toyon	/	/	toyon

	Serinsim	Takutan	Poring	Kiau
Bundu Tuhan	6	10	8	10
Kiau	6	10	8	
Poring	8	10		
Takutan	8			

Figure 7.2 A comparison of the number of cognate plant names between each pair of communities for various species or groups of species of non-rattan palms.

region are called **terintid** in Kiau and Bundu Tuhan, but **begung** in the other communities.

You can use results such as those presented in Table 7.1 to compare the similarity between the way that different communities classify plants and other objects. Simply count the number of matches between each pair of communities. The results can be presented in a matrix such as the one shown in Figure 7.2, which shows the number of cognate plant names between each pair of communities for the various species or groups of species of non-rattan palms listed in Table 7.1. Note that Serinsim, which is the most isolated of the five communities, has relatively fewer matches than the other localities.

An analysis of the similarity of classification in various communities is not just a theoretical exercise, but may be applied in the preparation of educational materials about local knowledge of plants. Before writing and distributing booklets on the use and naming of the regional flora, you should verify if all communities label the species in the same way. If variation does exist, take care to note the correct name for each community. Similarly, differences in use and management of plants should be included. For example, state that 'Oncosperma horrida is called **melugus** in Poring and Serinsim, but the residents of Takutan refer to it as **telibung**. The name **telibung** is also used in Kiau and Bundu Tuhan but the palm apparently does not grow in these higher elevation communities.'

Although identifying cognate terms may appear to be easy, you should exercise caution in differentiating true cognates from terms which have been borrowed from another language or which are simply descriptive phrases which have arisen coincidentally in the various communities. In a discussion of plant nomenclature from two dialects of Maya in southern Mexico, Brent Berlin, an American anthropologist who has worked extensively in Mexico and Peru, and Charles Laughlin, a linguist from the Smithsonian Institution, set out guidelines for judging true cognates [6]. They found some pairs of names that are exactly the same and others which show minor phonological variations which are common in many other word pairs from the two dialects. For example, 'pine' is called **tah** in Tzeltal

Maya and **toh** in Tzotzil Maya, and 'beggar ticks', a herb with barbed fruits, is named **mahtas** in Tzeltal and **matas** in Tzotzil. An 'a' in Tzeltal which corresponds to the 'o' in Tzotzil and an 'h' in Tzeltal which is not present in the same phonological context in Tzotzil are common differences found in many other instances. Linguists refer to this type of systematic difference as a **regular phonemic correspondence**.

Berlin and Laughlin discovered some names which are very similar, but which cannot be explained by a regular phonemic correspondence. For example, the 'dogwood tree' is called **siban** in Tzeltal and **'isbon** in Tzotzil. Although there is a phonological rule that would allow us to explain that the final 'a' in Tzeltal corresponds to an 'o' in Tzotzil, there is no reason to predict that the initial 'si' in Tzeltal would be a ''is' in Tzotzil. You may consider as cognates the names which show this type of skewed or infrequent phonemic correspondence, but linguists may ask you to offer more rigorous proof that these forms are in fact related.

Other pairs of plant names may be similar, but are likely to have arisen from a parallel way of describing a plant in different communities or from borrowing from another language. Berlin and Laughlin note that compound names such as Tzeltal **yašal ni wamal** and Tzotzil **yašal nič c'i'lel**, both translated as 'blue-flowered plant', should not be included as cognates. They also excluded other compound names which are linked by a single constituent such as Tzeltal **cacames** and Tzotzil **sat mes**. They analyzed separately 64 pairs of similar names containing Spanish loan words which have been introduced some time over the last 500 years. Although interesting as indicators of how each dialect has been affected by contact with Spanish speakers, they cannot be considered as true cognates derived from a single ancestral term in the protolanguage.

In the end, Berlin and Laughlin were able to discover 111 sets of plant names which they considered to be cognate expressions. By characterizing each folk category according to a four-point scale – (1) cultivated, (2) protected, (3) useful wild and (4) wild insignificant – they were able to find a correlation between cultural significance of plants and their tendency to be labelled by cognate names. 87% of the cultivated species in the sample had similar names in the two communities but this level dropped with decreasing cultural importance – 80% of protected plants, 45% of wild useful plants and 17% of wild insignificant plants were called by names considered to be cognates.

7.5.3 Analyzing the structure of plant and animal names

Although folk botanical nomenclature is not guided by any set of written rules, there are striking similarities in the way that plants are named by local people around the world [115–117]. The technical terms used to describe the structure of folk names have become well established in the scientific literature, but they may be difficult to understand for researchers without a linguistic background. A basic step in analyzing the structure of folk botanical nomenclature is to grasp the difference between primary and secondary names and to distinguish between the various sorts of primary names.

A **primary name** is considered to be 'semantically unitary', which means that it is a single expression, even if composed of more than one constituent. In fact, many primary names have just a single constituent, such as the English bird names *eagle* or *hawk*. Other primary names are composed of more than one constituent, yet they function in language as a simple name. For example, in English we think of (and write) *bluebird* as a single word, even though it is composed of two easily recognizable constituents – 'blue' and 'bird'. Primary names are thus said to be unitary, because people say, write and think of them as single words.

Secondary names are formed from primary names simply by adding a modifier which further describes the plant or animal. In English, for instance, we speak of different types of *eagles* – the *bald eagle*, the *golden eagle*, and so on. In each of these cases, a primary name – 'eagle' – is made into a secondary name by adding a modifier, such as 'bald' or 'golden'.

In practice, primary names can be distinguished from secondary names in a variety of ways. We can recognize some primary names right away because they are composed of a single constituent. But how do we tell apart compound names such as bluebird (a primary name) and blue finch (a secondary name)? There is no *a priori* linguistic distinction between the terms, but the difference often becomes apparent when we hear the names used in everyday speech and when we ask about folk categories in diverse social contexts.

True names can be distinguished from descriptive phrases in a similar way. In English, for example, we can talk of 'bluebirds' and 'blue birds', meaning in one case a particular species and in the other instance all birds which are blue. When studying another language, we encounter similar situations – is a term which translates as 'red flower' the name of a folk category or just a description of all plants which have red flowers? In the course of conversations and interactions with native speakers, it is important to discover which terms are true names of folk categories and which are merely descriptive.

As noted above, there are several types of primary names. In English, for example, *oak* is the general name for a class of hardwood trees, *crabgrass* is the general American name for several genera of weedy grasses, and *redwood* is a general name for a conifer that grows on the west coast of the United States. All three are primary names, but the structure of each is different. Oak is **simple**, composed of a single word that we cannot break down further. The other names are said to be **complex**, because we can break down crabgrass into 'crab' + 'grass' and redwood into 'red' + 'wood'.

Note that crabgrass includes the name of 'grass', the higher category or lifeform to which the plant belongs. This type of name is called a **productive complex primary name**. Redwood does not include the name of 'tree', the lifeform in which it is included. A general name of this sort is called an **unproductive complex**

simple (composed of a single constituent)

Primary name (semantically unitary)

productive (higher category named)

Lexeme   **complex** (composed of two or more constituents)

unproductive (higher category not named)

Secondary name (semantically binary expression; higher category named; found in contrast set of categories with similar names)

Figure 7.3 A diagram of the different types of plant names with key characteristics mentioned in parentheses.

primary name. A general scheme which summarizes these types of names is presented in Figure 7.3.

In most indigenous languages, we find numerous examples of each type of primary name. Table 7.2 gives several Mixe examples drawn from various lifeforms.

As noted above, plant names that are formed of a primary name and a modifier are called **secondary names**. These names are composed of a primary name that indicates the higher category in which the plant is included, plus one or more words that describe the plant in some way. For example, English speakers have several specific names for types of oaks: *pin oak*, **red oak**, **white oak**. Each of these names is composed of a primary name (oak), and a modifier (pin, red, white). Categories labelled by secondary names always belong to sets having two or more members, each contrasting with the others.

Folk names, unlike scientific names, often reveal information about the appearance, utility, or distribution of the plant. This is particularly true for the names of

Table 7.2 Examples of the three types of primary plant names taken from four Mixe botanical lifeforms

Lifeform	Type of primary name		
	Simple	Productive complex	Unproductive complex
kup	*xijts*	*alivia kup*	*iitsum tsi'ik*
tree	*avocado*	*relief tree*	*peccary musk*
ojts	*ko'on*	*jan ojts*	*kaaj aaxk*
herb	*tomato*	*fever herb*	*animal tick*
aa'ts	*ejkx*	*aya'ax aa'ts*	*eex taats*
vine	*chayote*	*cry-baby vine*	*crab teeth*
tsoots	*veek*	*tsookun tsoots*	*pa'a peetun*
grass	*spikerush*	*rain-cover grass*	*pathside broom*

specific and varietal categories, which typically contain a modifier that alludes to the color, origin or other property of the plant. Mentioning these features probably helps people to remember which names go with which plant.

As you transcribe and analyze the meaning of names, you can also begin to classify them as simple primary, complex primary or secondary names. This will contribute to your understanding of the folk categories you are studying. Keep in mind that simple primary names are typically associated with plants that are morphologically distinctive or culturally important. Productive complex names encode information on the lifeform to which the category belongs and both types of complex names may provide information about the morphology, use or symbolism of the plant. Secondary names, because they occur in contrast sets, can often lead you to discover other specific and varietal categories of which you had not been aware.

There is an interesting correlation between folk categorization and nomenclature. As shown in Table 7.3 (p. 215), lifeform and generic names are primary names, while specific and varietal categories are usually labelled by secondary names. The kingdom and any intermediate categories are typically unnamed.

7.6 Free listing

One of the first steps in gathering linguistic data for ethnobiological studies is to obtain a list of local terms that refer to the topic that we are researching. In other words, we define and delimit the **domain** – the subject that interests us – and discover how local people talk about it. A rapid, participatory approach is to ask community members to list the names of categories that belong to the domain [58]. For example, if you are interested in the types of firewood being used by Mixe people, you would first determine that there is a native category for 'firewood', referred to as *ja'axy* in Totontepec, which corresponds to woody plants used as fuel. Then you would ask community members in their local language, 'What kinds of *ja'axy* do you know?'

This technique, called **free listing**, helps us to understand if the domain is **salient**, which means that it is considered culturally important and easily recognizable by the people we are interviewing. Imagine if you asked desert-dwelling Amerindian people for a list of cactus names and an Amazonian Indian group for a list of vines. You would come up with fairly long lists. Switch the questions and you would discover that you were posing the wrong question for that particular culture. Not only would there be a lack of generic plant names for these lifeforms, but you would probably have trouble posing the question, because the tropical forest-dwellers would lack a general term for 'cactus' and the desert people a name for 'vine'. Because not all cases are this obvious, it is best to test any question with four or five people in each community or social group to see if they respond in a consistent way which demonstrates they understand what it is that you are requesting.

When we ask the right question, free listing can give us not only a fairly complete set of native categories, but also information on which are the most culturally important. When people are asked to freely recall things, they tend to list the most significant ones first. In addition, prominent categories are cited by almost everybody, while less significant ones are mentioned by a minority of informants. These observations can be easily quantified. The number of people who mention the category is simply totaled. The average order in which each category is mentioned can be calculated by adding together the order in which each respondent mentioned the category and dividing by the total numbers of respondents.

When we consider these two variables together we can calculate an **index of relative saliency**, a ranking of the most to least important categories. This index can be used to decide which plants will be used in subsequent eliciting tasks and it can make us aware of important categories which have not yet been mentioned in open interviews. The best way to visualize these results without carrying out extensive calculations is to prepare a graph on which the x-axis corresponds to the number of people who mention a given category and the y-axis corresponds to the average order in which the category is mentioned. If there is a relatively high correlation between these two variables, you will find that the categories in the bottom right hand part of the graph are more salient than those in the upper left hand part.

How many people should be asked to give a free list? The answer depends on the size of the domain. If it contains only 50 categories, you need fewer respondents than for a domain of 500. For medium-sized domains (less than 100 or so total categories), it is a good idea to begin the inquiry with approximately 20 to 30 people. Once you begin to observe that most of the responses given by new informants are only repeats of those previously given, you can assume that enough people have participated.

If you are working with a domain that contains more than a hundred categories, it becomes very time-consuming to ask many people to give a complete listing. For example, it is impractical to ask a large sample of informants for the names of all folk botanical generics, because any knowledgeable person is likely to know several hundred. Brent Berlin and his colleagues elicited complete inventories from 13 Tzeltal-speaking informants during the course of their research in Chiapas, Mexico. The resulting lists contained from 187 to 565 names and the average number per informant was 398. It would be difficult to collect and analyze similar data from a large number of informants. Even with 30 informants, there would be approximately 12 000 responses to transcribe and analyze.

In the case of large domains, you may wish to ask each respondent to name a limited number of categories rather than providing a whole list. In Totontepec, we asked 150 adults to provide the names of 25 useful plants. The resulting list, which contained nearly 300 folk generics, allowed us to judge the completeness of the ethnobotanical inventory that we were carrying out in the community.

7.7 Systematic surveys of local plant knowledge

Collecting systematic lists of plant names leads quite naturally to exploring folk classification, a subject which has been of major interest to linguistic anthropologists over the last three decades. After you have obtained a list of plants or animals, either through a free listing task or by making an ethnobiological inventory, you may wish to analyze the structure of the folk categories and their indigenous names. Although time-consuming, an in-depth study of folk classification is important if you are seeking to promote the survival of local systems of biological knowledge, particularly in communities undergoing rapid culture change. Before you seek to return the results of your research in the form of popular booklets on plant use or bilingual dictionaries that describe biological organisms, you should understand the basic structure of the local system of ethnobiological classification and explore how it corresponds to scientific taxonomy.

7.8 Categories of ethnobiological classification

Although many ethnobiologists have proposed ways of analyzing local people's categorization of plants and animals, a scheme of folk classification devised by Brent Berlin is gaining general acceptance among researchers [113, 116]. It has proved useful for comparing the biological categories of local people from many different world cultures. Table 7.3 gives a synopsis of the various ranks of folk botanical categories in this scheme, using Mixe classification of the scarlet runner bean as an example of how the categories at each rank are named.

The most general category is the plant **kingdom**, which is implicitly recognized by local people and is often contrasted with animals. It contains all higher plants and may also include mosses, lichens and other similar organisms. **Lifeforms** are broad, distinctive classes – such as trees, vines or grasses – recognized by their habit, their distribution in a specific ecological zone, their utility or by a combination of these factors.

Table 7.3 A summary of Brent Berlin's scheme of folk classification, employing Mixe categorization of the scarlet runner bean, *Phaseolus coccineus* L. as an example. NA, not applicable.

Rank	Type of name	No. of members in one system	Mixe name and English gloss
Kingdom	Often unnamed	1	Unnamed
Lifeform	Primary name	5–10	*aa'ts* vine
Intermediate	Often unnamed	NA	*xajk* bean
Generic	Primary name	300–600	*maj_xajk* large bean
Specific	Usually secondary name	Typically, some 20% of generics contain specifics	*yak maj_xajk* black large bean
Varietal	Secondary name	Relatively few varietals exist	NA

Intermediates are small groupings of several generics that are perceived to be similar in some way. These groupings are called intermediates because they are in-between lifeforms and generics, the most important categories in the folk hierarchy. Intermediates are also called **covert categories**, because they often go undetected in folk classifications, are usually unnamed and may not be recognized by everyone in the community. For example, the oak tree genus, which is very diverse in the northern Sierra of Oaxaca, is divided into several named folk generics by the people of Comaltepec. Although most Chinantec-speakers realize that all oaks belong to a single class of trees, they do not give a name to this overall category – it is an intermediate in Chinantec botanical classification.

Generics are the most salient categories in folk botany. They are the first to be learned by children and are frequently mentioned in interviews, surveys and general conversation. This is in part because they are numerous and easy to recognize. Each one forms a morphologically distinct unit, often found in a certain sector of the community and distinguished by its cultural significance.

Most generics are included in – or **affiliated** to – a lifeform, although some morphologically distinct or economically important plants may be **unaffiliated**, or independent of all lifeforms. For example, many edible beans are included in the Mixe lifeform that corresponds to vine, but corn is unaffiliated; it is not considered to be an herb, grass, tree or part of any other lifeform. This is probably because corn is of primary importance in the diet and economy of the local people and because it looks very different from other plants grown in the Sierra Norte.

Some generics are further divided into **specific** categories. In English, as in many languages, the generic that corresponds to 'oaks' is divided into many specifics that are called 'pin oak', 'black oak', 'burr oak' and so on. The Chinantec recognize many different specific types of corn, which are given names such as 'black corn', 'white corn' and 'red corn'. These specifics are distinguished by the color of the dried kernels, the geographical locality in which the plants flourish, the length of time it takes the cobs to mature and other features.

Very rarely, specifics are partitioned into **varietals**. For example, the Mixe have a specific corn category called 'ancient corn', which they further subdivide into varietals known as 'yellow ancient corn', 'white ancient corn', 'large ancient corn', 'red ancient corn' and 'black ancient corn'.

In general, specifics and varietals are differentiated by a few morphological characters such as coloration, size or shape of plant parts. They may also be linked to a particular use or microclimate in the community. Specifics and varietals often correspond to agricultural or other culturally significant plants. They are usually more prevalent in the folk classifications of traditional agriculturalist societies than among nomadic people who depend largely on foraging and hunting for survival.

The folk botanical classifications of traditional cultivators from around the world contain roughly the same numbers of categories at each rank. Typical examples from Totontepec Mixe and Comaltepec Chinantec botanical classification

Table 7.4 The approximate number of categories at each ethnobotanical rank in Comaltepec Chinantec and Totontepec Mixe botanical classifications

Ethnobotanical rank	Number of Chinantec categories	Number of Mixe categories
Kingdom	Unitary	Unitary
Lifeform	11	10
Intermediate	Many	Many
Generic	388	416
Specific	306	394
Varietal	6	20

are summarized in Table 7.4. There is a single kingdom, several lifeforms and between 300 and 600 generics in a typical system. Approximately 20% of the generics are further subdivided into specific categories, that tend to number from 2 to 10 per generic.

The total number of specifics varies widely. There may be hundreds in a folk biological classification, but they rarely surpass the number of generic categories. Very few specifics are further split into varietals, which are relatively rare in folk botanical classification. Intermediate categories have been insufficiently documented to permit an estimate of their relative frequency.

7.8.1 Cross-cutting categories and parallel classifications

Some plant categories used by local people do not fit into the general system of folk classification described above. For example, the Chinantec have words for fruit ($'o^L h u \ddot{\iota} \ddot{\iota}^L$) and for firewood ($k u \ddot{\iota} \ddot{\iota}^{LH}$), the Mixe have terms for edible greens ($t s \underline{u}' \underline{u} p$) and for medicinal plants ($t s \underline{o o} j y$). These named categories are delimited primarily by their use and do not correspond to lifeforms, generics or categories of any other folk rank. These types of groupings are called **cross-cutting categories**, because they often include plants from several different lifeforms, thus cutting across the general-purpose classification of plants.

Some cross-cutting classifications are based on other criteria than use. Most ethnic groups of Mexico, for instance, classify plants according to their **humoral property**, the perceived quality of being hot, temperate or cold in their effect on the human body. This hot–cold system of classification (which some people think is indigenous and others believe was introduced by the Spanish) is applied only to plants which are consumed, either as medicine or food. Plants with large quantities of essential oils, such as mints and mustards, tend to be considered as hot. Succulent plants, such as begonias and species of *Crassulaceae*, are usually thought of as cold. Humoral property is not correlated with the division of plants into lifeforms. Some herbs might be considered as hot and others as cold and the same holds true for trees, vines and other broad folk taxa.

At times, these systems of classification are extensive and complex, providing an alternative way of ordering the majority of plants known by the people of a certain culture, in which case they are called **parallel classifications**. Jacques Tournon, a French biophysicist who carries out ethnobiological research in the Peruvian Amazon, found that the Shipibo-Conibo classified many plants in two ways – as members of folk generic categories and as *rao*, a category of plants, animals and minerals that are classified by the way they affect human health and behavior. Depending on the context in which they are discussing the plants, the Shipibo-Conibo use either the generic name or the *rao* name, or sometimes both. For example, *Hura crepitans* L., a spiny tree with copious toxic latex, is usually called by the generic name *anà*, but is also referred to as *peque rao* in the context of its use as a treatment for persistent skin wounds caused by leishmaniasis, a tropical disease.

Ethnobotanists find that exploring alternative systems of classification is helpful in some research projects. For example, if we are interested in the fuel sources of a community, we will want to determine if a general category for firewood exists and how it is structured. No matter what the scope of the research, it should be clearly defined which categories constitute part of the general folk classification, which belong to alternative, special-purpose ways of ordering the plant world and how the two systems are interrelated.

7.9 The correspondence between folk and scientific classification

Charles O. Frake, an American linguistic anthropologist who focused on how people perceive and classify the world, described how ethnographers usually go about collecting linguistic data [117]:

> A relatively simple task commonly performed by ethnographers is that of getting names for things. The ethnographer typically performs this task by pointing to or holding up the apparent constituent objects of an event he is describing, eliciting the native names for the objects and then matching each native name with the investigator's own word for the object. The logic of the operation is: if the informant calls object X a *mbubu* and I call object X a *rock*, then *mbubu* means *rock*. In this way are compiled the ordinary ethnobotanical monographs with their lists of matched native and scientific names for plant specimens.

Frake and many other ethnoscientists observed that the correspondence between folk and biological categories is not this simple. In the remainder of this chapter, I will describe some of the techniques that allow researchers to gain a clear understanding of the breadth of local biological categories and the correspondence between folk and scientific classifications of the natural environment.

Researchers have discovered a clear relationship between scientific and folk botanical classification in many different cultures. As Eugene Hunn remarked after studying folk classification in southern Mexico [118], '... local people demonstrate an intimate and empirically reliable knowledge of the local flora and fauna and share with the field worker an appreciation of the ordered complexity of the living world'.

How can this shared knowledge be characterized and measured? The simplest way is to describe the correspondence between folk categories and scientific categories. This approach is most successful for generic and subgeneric taxa, which often correspond to a scientific species, group of species or a genus. The most common way of characterizing correspondence is to show how a single folk generic **maps onto** scientific species, as shown in Figure 7.4.

Generics often show a **one-to-one correspondence** with scientific species. For example, the Mixe generic for dogwood exclusively corresponds to *Cornus disciflora*, the only species of this genus that grows in Totontepec.

Species of great cultural significance are **over-differentiated**, or split into many distinct categories by local people. The avocado, *Persea americana*, is partitioned into two generics and several specifics by the Mixe; *kooydum* pertains to large-fruited avocados similar to the ones sold grown and commercialized worldwide, whereas *xijts* corresponds to a local variety of the avocado tree which bears small thin-skinned fruits that have an anise-like taste and smell.

Species that are less important culturally or less distinctive in appearance are usually **under-differentiated** – they are lumped into a single generic. Several species of *Inga*, all used for firewood, are included in the Mixe generic *ii'k*. Two species of *Inga* with particularly large fruits containing edible pulp are placed in the folk generic *ta'chki*.

When speaking of correspondence, it is essential to define which folk and scientific categories are being compared. As noted above, most comparisons are between folk generics and botanical species. This is illustrated in more detail in Box 7.1.

In addition, it is possible to compare any folk rank with any scientific rank. As shown in Figure 7.4, we can say that the Mixe over-differentiate *Inga* at the folk generic rank – they divide it into two distinct classes. And while *Cornus* has a one-to-one relationship with the Mixe generic *tsaa'k*, it is over-differentiated by the Mixe at the specific rank, where it is divided into two classes.

At higher ranks, the relationship between folk and scientific taxa is less exact than at the generic or subgeneric level. The content of kingdom is generally similar in the two systems – local people and scientists disagree little on what constitutes a plant. The delimitation of intermediate categories often agrees quite well with the circumscription of a botanical family. Lifeforms do not always correspond to botanical taxa, but they make sense to botanists because they often correlate well with descriptions of plant habit – tree, vine, herb and so on. A few

	Scientific species	Folk generic	Folk specific
E			**poop'p tsaa'k** (white dogwood)
Q			
U	*Cornus disciflora* DC.	**tsaa'k** (dogwood)	
A			
L			**yak tsaa'k** (black dogwood)
O			**kooydum** (typical avocado)
V			
E			
R		**kooydum** (avocado)	
-			
D			**pa'ajk kooydum** (sugary avocado)
I			
F			
F			
E	*Persea americana* Miller		
R			
E			**poo'p xijts** (white avocado)
N			
T			
I		**xijts** (avocado)	
A			**tsapts xijts** (red avocado)
T			
E			
D			
U			
N			
D			
E	*Inga latibracteata* Harms		**poo'p ii'k** (white *Inga*)
R			
-			
D	*I. oerstediana* Benth.	**ii'k** (small-fruited *Inga*)	
I			
F	*I. schiedeana* Steudel		**tsapts ii'k** (red *Inga*)
F			
E	Other *Inga* species		
R			
E	— — — — — — — — — — — — — —		
N			
T	*Inga jinicuil* Schldl.	**ta'chki** (large-fruited *Inga*)	
I			
A	*Inga paterna* Harms		
T			
E			
D			

Figure 7.4 Mixe examples of the mapping of scientific species to folk generics and specifics, showing cases of one-to-one (equal) correspondence, overdifferentiation and under-differentiation at the generic rank.

Box 7.1 The correspondence of Tzeltal Maya plant generics to botanical species

As part of their analysis of Tzeltal Maya folk botany, Brent Berlin and his colleagues looked at the correspondence between plant generics and botanical species in a systematic way [7]. The results of their analysis are shown in Table 7.5.

Table 7.5 Correspondence of Tzeltal plant generics to local botanical species

Type of correspondence	Number of generics	Percentage of generics
One-to-one	291	61
Under-differentiation, type 1	98	21
Under-differentiation, type 2	65	14
Over-differentiation	17	4
Total	$n = 471$	100

They found that most generic categories included one and only one scientific species, which is called one-to-one correspondence. There were very few cases of over-differentiation, that is, when two or more folk generics correspond to a single scientific species.

Finally, there was under-differentiation in over one-third of the cases. Berlin considered two types of under-differentiation – when a generic refers to two or more species of (1) the same genus or (2) more than one scientific genus. As an example of the first type, he gives the category **ch'ilwet**, which refers to some five species of *Lantana*, a genus in the *Verbenaceae*. The second is exemplified by **tah**, a folk generic which includes several species of *Pinus* and at least one species of *Abies*, both genera of the *Pinaceae*.

lifeforms, such as 'grass', 'fern' or 'palm', correspond to a botanical family or higher category.

As you analyze folk classification in the community in which you are working, ask people about the characteristics which differentiate the various categories. Pay particular attention to variations in plant morphology, uses and habitat and how they affect the correspondence between folk and scientific categories.

8

Ethnobotany, conservation and community development

Figure 8.1 Danson Kandeung felling a stem of *Eugeissona utilis* Becc. near Kinabalu Park. This species, an endemic palm of Borneo which is a source of the traditional food **sago**, has multiple stems which sprout from the base of the plant. Felling one trunk allows people to harvest the starchy pith, but does not kill the palm. Reconciliation between protection and utilization of biological organisms is a key element of conservation and development projects.

8.1 Applying traditional ecological knowledge

If you have read this manual from the front cover to here, you have learned about field methods in ethnobotany and have explored the relationship between folk and scientific knowledge of plants. Recalling that the manual forms part of a series on plant conservation, you might ask yourself, 'How can I employ the results of ethnobotanical research to strengthen the protection of natural areas and contribute to the well-being of local communities?'.

This question is on the minds of many people who work in the field of conservation and development [119–121]. Since the birth of ethnobotany as an academic discipline, researchers have pointed to the many benefits that would result from studying the ecological knowledge of forest dwellers, traditional agriculturalists, pastoralists and other people living close to nature [122–126]. These voices have been heeded and there is now a general consensus that local people have an important role to play in any project that seeks to discover the secrets of the forest, assess its economic value and conserve it [127–131].

8.2 Ethnobotanical research and community development

Ethnobotanists have traditionally directed their efforts towards one of two goals. Economic botanists sought to discover new natural products of commercial value, often for the benefit of the developed world, whereas ethnoscientists focused on achieving a theoretical understanding of how people perceive and manage the environment. Since the late 1960s, many ethnobotanists, building upon and modifying these early goals, have directed their attention to applying the results of their research to conservation and development problems [132].

The community projects in which they participate have various goals, including return of the research results to host communities, strengthening traditional systems of agricultural production, encouraging rational use of plants in health care and promoting traditional ecological knowledge. Although the scale and objectives vary for each project, all are driven by the enthusiasm and idealism of the researchers, technicians and residents who are working together to improve local conditions [133].

8.2.1 Community projects

A complementary manual in this series (*People and Wild Plant Use* by Anthony B. Cunningham) describes in detail how to apply research results to conservation and development. What I offer here is no substitute, but rather a collage of projects that can provide inspiration to colleagues who are struggling with this vital issue. Because I was able to visit only a few of the projects cited below, I have sketched the descriptions from written accounts or conversations with key participants. As such, my impressions probably reflect better the aspirations of the organizers rather than an up-to-date assessment of their accomplishments. In providing these synopses, my intention is present a few ideas that may be experimented with in different cultural and ecological situations.

The topics are grouped under various broad titles, including forests, conservation of wild crop relatives and endangered useful plants, education and protected areas. They range from specific techniques that ethnobotanists can employ as they carry out their research – such as building local herbaria or promoting exhibits of useful plants – to major initiatives, on the order of extractive reserves and community health programs, in which they can serve as consultants.

8.3 Forests

8.3.1 Agroforestry and agroecosystems

Agroecology focuses on the complex ecological relationships that are at the foundation of any system of agricultural production [134, 135]. These include the interaction between crop plants and insects, soil microorganisms, weeds and many other elements of the local environment. Ethnobotanists can work alongside agroecologists to document traditional systems of agricultural production and to assist in transferring appropriate technology from one region or ethnic group to another. The ultimate goal of these efforts is to design agroecosytems that blend ecological concepts of integrated pest control and organic fertilization with elements of traditional polyculture – the cultivation of several species of crop plants in a single plot, often intermixed with semi-cultivated plants [136].

Ethnoecological research may focus on indigenous systems of soil classification, local methods of insect control or the input of non-cultivated plants in agroecosystems, and has contributed to the general recognition of the value of polycultures. Ethnobotanists can play a catalytic role in suggesting which wild or semi-cultivated species can be incorporated into agroforestry or agroecosystems [137]. They can also propose alternatives to environmentally destructive practices such as large-scale plantation agriculture or cattle-raising. Many of these alternatives have been inspired by ancient systems of cultivation that are only now being rediscovered, or by traditional systems that are finally being given the recognition they deserve [138]. In the Beni region of Bolivia, for example, archaeologists and agronomists are working together to design and test modern-day versions of *camellones*, or raised beds, that once permitted intensive agricultural production in the marshy savannahs of the region. Their hope is to provide an alternative to the swidden agriculture that is destroying forests in the Beni Biosphere Reserve. In a parallel project, an assessment of the plants used by Mestizo immigrants and Chimane Indians who live in and around the Biosphere Reserve is documenting some of the botanical resources that can serve as inputs to these agroecosystems.

8.3.2 Social forestry

Social forestry is a broad term used to describe initiatives in which communities act as the primary managers of forested lands. Based on their often detailed knowledge of the natural environment, local people are taking the lead in reforestation, selective logging, ecological reconstruction of fragile lands and sustainable harvesting of non-timber forest products. Many of these projects are carried out in collaboration with governmental forest department (joint forest management). Ethnobotanists can participate in these community projects by documenting how local people's management of the forest favors the growth of native species which have subsistence and commercial value. Native woody plants often act as nurse trees, providing protection for herbs and shrubs, many of which are used by local people. Regeneration of the natural forest reduces the need to reforest with exotic species which are of limited usefulness and which often have negative impacts on the local environment, such as inhibiting undergrowth plants.

In many parts of India, there is a wealth of experience in social forestry. At a workshop on sustainable forestry held in New Delhi in 1990, participants reviewed the joint management of eight state systems of forest reserves, ranging in size from 5000 ha to over 200 000 ha. In all of the projects, the forest department recognizes the right of people in forest protection committees to gather minor forest products, including fodder, fibers, firewood and construction materials as well as edible and medicinal plants. In some cases they are also allowed to harvest a limited amount of timber. In return, the local people are expected to protect the forest, including taking action to limit destructive activities such as grazing and unauthorized timber harvesting. They also take part in management activities, including enrichment planting, that is the selection, planting and care of particularly valuable species which are cultured in gaps in degraded forests. The forest departments often reciprocate by paying wages to local laborers and improving processing and marketing capabilities which allow communities to capture a greater share of the profits earned from minor forest products. These joint forest management programs are promoted as one solution to reverse the rapid rate of deforestation that India has experienced in recent decades.

8.3.3 Marketing of non-timber forest products

At the end of a seminar on the sustainable harvesting and marketing of rain forest products sponsored by Conservation International in Panama in 1991, participants issued a declaration which urged ethnoecologists, development agencies and conservation organizations to pay greater attention to the commercialization of forest products [97]. In particular, they encouraged investment in this trade and revision of governmental policies that restrict the flow of economic benefits to local people.

In order for this approach to be successful over the long term, they recognized the need to avoid the ecological damage or rapid cultural change that is often associated with exploitation of the forest [139]. They envisioned joint ventures between conservation organizations, governmental agencies and businesses that would link commercialization of forest resources with biodiversity conservation and community development. Gary Nabhan [140], one of the participants in the seminar, offered some guidelines about linking community development and conservation through applied ethnoecological studies (Box 8.1).

Box 8.1 Guidelines for linking applied ethnoecology to conservation and community development

Gary Nabhan, an ethnobotanist at the Desert Botanical Garden in Arizona, has extensive experience in working with indigenous people in Mexico and the southwestern United States. Among other initiatives, he has promoted the preservation of land races of crop plants of the region, many of them cultivated by fewer and fewer traditional agriculturalists. Upon reviewing the impact of development and conservation initiatives based in part on traditional ecological knowledge, he developed guidelines that ethnoecologists should consider before embarking on applied projects. In his own words, these guidelines seek to ensure 'that indigenous peoples and other peasant communities benefit from applied ethnobotanical development and that projects sustain rather than deplete biodiversity',

(1) The project should attempt to improve the objective and subjective well-being of local communities – the latter as members perceive themselves – rather than simply seeking cheap production sites and importing inexpensive labor.

(2) Cultivation in fields or through agroforestry management should be considered if there is the threat that wild harvests will deplete the resource in its habitat. At the same time, the new production should offer a new source of employment to those who were formerly dependent on the wild harvest.

(3) Wildland management and more sensitive harvesting practices should be introduced if the resource might sustain economic levels of extraction in the habitat, eliminating the need to convert that habitat to conventional agriculture or livestock production. However, the biotic community should be monitored to determine whether the regenerative capacities of the targeted resource or its associated species are measurably affected.

(4) The plant(s) chosen should offer multiple products or be adapted to diversified production systems, so that small producers are buffered from the vagaries of any single economic market.

(5) Whenever possible, the project should build on local familiarity, use and conservation traditions for the plant being developed.

(6) If possible, the economic botany project should draw on locally available genetic resources, technologies and social organizations, so that local people can retain control over the destiny of the resource.

An agenda similar to the Panama declaration has been proposed by researchers who are active in supporting **extractive reserves**, natural areas set aside for harvesting non-timber forest products which are commercialized in regional or international markets [141–143]. Although extractivism began as a highly exploitative practice based on the near enslavement of rural workers (88,104,105), some development specialists feel it can be modified to become a model of environmentally sound development in the tropics. In the Brazilian state of Acre, for example, the National Union of Rubber Tappers (*Conselho Nacional dos Seringueiros*) has established a network of extractive reserves in which middlemen are being replaced by producers' cooperatives, and health and educational facilities are provided for the rubber tappers and their families.

Several organizations are active in supporting the marketing of products from tropical forests of Latin America and other areas. The Technology Foundation of the State of Acre, with support from the International Tropical Timber Organization (ITTO), has analyzed the marketing of medicinal plants and Brazil nuts harvested from extractive reserves in the state of Acre. Conservation International has promoted the production and commercialization of vegetable ivory, a product derived from tagua nut palms (*Phytelephas* spp.), as a way of supporting community development in selected communities in the Esmeraldas province of Ecuador. Cultural Survival Enterprises, created in 1989 to test alternative ways of generating income from the rain forest, has developed a broad portfolio of non-timber forest products which are sold in the United States and Europe. The primary goal of this initiative is to develop and expand markets for these goods, thus encouraging communities to engage in extractive activities which are sustainable and profitable.

Coordinators of the Periwinkle Project of the Rainforest Alliance, an American conservation organization, are focusing their research on international and domestic markets for pharmaceutical products derived from tropical plants. They are planning to return the results of this marketing research to communities and researchers – particularly in Brazil, Ghana and Indonesia – who are looking for ways to increase local incomes while ensuring the sustainable use of forest resources. Since 1991 they have been collaborating with Gemima Cabral Born, a Brazilian ethnopharmacologist who is working in the Atlantic Forest, a highly endangered vegetation type that extends along the Brazilian coast. As part of her ethnopharmacological survey of plants used by Caiçaras communities in the Jureia Itatins Ecological Reserve, she is promoting the sustainable management and commercialization of some medicinal plants.

8.4 Conservation of wild crop relatives and endangered useful plants

Conservation areas around the globe provide environmental services and harbor biological organisms important to local communities and to the world as a whole. In the past, few people recognized the urgency of preserving species that had no known economic value. With recent advances in crop plant genetics and the surging interest in biotechnology, many scientists increasingly recognize that wild species can play a vital role in future economic development [144].

Many examples from ethnobotanical exploration are strengthening this perspective, generating additional public support for biodiversity conservation. In the late 1970s a group of American and Mexican botanists working in Jalisco state discovered *Zea diploperennis*, a relative of maize which is a promising source of genes for improving disease resistance and other qualities of corn. In 1987 the Mexican government created the 140 000 hectare Sierra de Manantlán Biosphere Reserve in order to protect populations of *Z. diploperennis*, which cover about 360 hectares, as well as the surrounding flora and fauna. Bruce Benz, an American ethnobotanist, is working with Mexican colleagues to document local knowledge of the biosphere flora, with the ultimate goal of contributing to the development of local communities and continued protection of the Sierra de Manantlán.

Protection of plants and animals should always be balanced against the needs of local communities to have access to the biological resources essential to their subsistence and commercial activities. In several biosphere reserves along the border of the United States and Mexico, a team of Mexican and American researchers headed by Humberto Suzan and Gary Nabhan is seeking to preserve the rarest species in these arid lands while encouraging traditional plant utilization by the Papago Indians and other residents of the region. The surveys they have carried out in Rancho del Cielo, Big Bend National Park and Organ Pipe Cactus National Monument reveal the important role that protected areas can play in the preservation of rare species. Over 65% of the rare plant species of northeastern Mexico were discovered in Rancho del Cielo alone. By focusing on endangered species which are currently being harvested by local people, they hope to find ways to fulfil both the conservation and development objectives of biosphere reserves.

8.4.1 Botanic gardens

Botanic gardens have a prominent place in the history of economic botany because they have been the point of introduction of many crop plants now cultivated around the world. Some gardens continue this legacy by growing wild relatives or landraces of crop plants and by experimenting with the cultivation of over-harvested wild species that could potentially be integrated into polycropping and agroforestry systems. The staff may also educate the public about the need to protect endangered habitats, provide training in techniques of plant conservation and carry out basic and applied research in natural areas around the world. Botanic gardens occasionally house facilities for analyzing the nutritional composition and pharmacological properties of plants.

Some ethnobotanists associated with botanic gardens participate in the identification and rescue of useful plants which are endangered by habitat destruction or over-harvesting. These species can be protected by conserving the remaining natural areas where they grow, an approach called *in situ* **conservation**. They can also be saved by collecting seed to store in germplasm banks or by propagating plants in botanical gardens, techniques which are referred to as *ex situ* **conservation**. As part of this conservation strategy, some gardens specialize in **re-introductions** – propagating endangered plants and transplanting them into wild areas that correspond to their original habitat.

Brien Meilleur, an ethnobotanist who formerly directed the Amy B.H. Greenwell Ethnobotanical Garden in Hawai'i, has called for regional botanic gardens to orient their efforts towards cultivation of local useful plants rather than exotic ornamental species, thus educating the public about the value of traditional ecological knowledge [145]. Other ethnobotanists have advocated that people in rural communities construct ethnobotanical nurseries where useful species can be cultivated. This not only provides a source of coveted medicinal and edible plants, but also serves to familiarize younger people with the herbs that are traditionally used in the community. The cultivated plots serve as demonstration gardens where over-harvested wild species can be brought into cultivation and eventually integrated into home gardens or managed forests.

The World Wildlife Fund (WWF), US, has been active in supporting a local ethnobotanical garden in the Sibundoy Valley of Colombia. It was established in 1988 by Pedro Juajibioy-Chindoy, a Kamsa Indian, to preserve local plants used traditionally by the Kamsa people, who inhabit marshlands and dense forests along the eastern slope of the Andes. Now covering more than 7 hectares of land outside the town of Sibundoy, the garden serves as a training ground for Kamsa children, who learn about horticulture and the traditional knowledge of their elders. The organizers hope to promote a series of satellite gardens in other Kamsa settlements which range from the Amazon lowlands up to mid-elevation slopes in the Andes, thus ensuring the preservation of a broad range of wild useful species and landraces of cultivated plants.

8.4.2 Community herbaria

Pressed and dried plant specimens deposited in a community center or school give local people a permanent record of the useful species documented in the course of a research project. These herbaria also allow local participants to continue research on their own, using the mounted plants when asking other community members to verify data about the names and uses of certain species. The specimens are stored in tightly sealed cabinets, protected from humidity, insects and excessive heat. If the plants are to be handled often by students or other community members, they are sealed within plastic envelopes. Such collections are particularly valuable if cross-referenced to popular publications that describe the uses of plants in detail.

Herbaria can be effective tools in working with the younger generation in communities. Apart from using prepared specimens in natural history classes, students can be asked to make herbaria of their own. Yolanda Betancourt, a biologist, and Irma Betancourt, an educator, held a medicinal plants workshop for children in Tlaxcala, Mexico in 1987 as part of a plant resources project at the Universidad Autónoma de Tlaxcala. They distributed a booklet that contained brief botanical descriptions of seven common medicinal herbs. Each child was provided with a line drawing of each species to be colored as realistically as possible, using live plants as models. They then pressed samples of the herbs, later pasting them and the line drawings on appropriate pages and filling out printed collection labels. Once the mini-herbaria were prepared, they were instructed to record the ways in which their mother, grandmother or other adult used the herbs. This approach not only introduced children to ethnobotanical methods, but also encouraged them to learn the knowledge of their elders, a traditional process that is breaking down in some communities.

8.5 Education

8.5.1 Educational programs for the young

Ethnobotanists are concerned about the rapid disappearance of traditional ecological knowledge. Culture change can occur in indigenous communities very quickly, particularly when a new generation takes advantage of opportunities not available to their elders, such as attending school, learning the national language and migrating to urban areas. Although these can be important steps towards raising the standard of living and literacy in rural areas, they often result in a loss of the traditional ways of interacting with the natural environment which equally yield many benefits. Ideally, children should value what they learn both at school and at home, becoming fully bilingual and bicultural if they so desire.

One way of accomplishing this goal is to provide opportunities for the younger generation to master the ecological knowledge of their elders and particularly that of the traditional doctors and specialist resource users in the community. Conservation International, a non-profit conservation organization based in the United States, has been supporting a program that seeks to reverse culture loss in specific communities in Costa Rica and Surinam. Traditional healers take on apprentices and children study local uses of plants as part of an overall effort to battle both deforestation and acculturation. In Conservation International's 1992 annual report, Mark Plotkin (an ethnobotanist who has worked primarily in Latin America) described the initiative: 'We had two goals when we devised this plan. First, we wanted to ensure that the ancient lore of rain forest plants is kept within the tribe. Second, we wanted to raise awareness among indigenous peoples that conservation can be a powerful means of preserving both their culture and their home.'

8.5.2 Newsletters

Newsletters are a simple and inexpensive way of disseminating information and exchanging opinions about conservation, traditional knowledge and community development. Usually consisting of a few pages of text and illustrations, they can be sent for the minimal postal rate to many members in a professional network. One example is the Botany 2000 – Asia newsletter, which was first issued in 1992 by the Regional Office of Science and Technology for South and Central Asia of UNESCO. It fosters communication between hundreds of Asian-based botanists who are carrying out research on floristic diversity and ethnobotany, among other subjects. It provides a forum for discussion of ethical issues and descriptions of new programs and meetings.

Some newsletters are intended not only for university-trained colleagues, but also for people living in rural communities. The Indigenous Food Plants Program (IFPP) of Kenya, started in 1988, seeks to distribute information on edible wild plants collected by researchers of the East African Museum over the last 90 years. In the words of Christine Kabuye, an IFPP coordinator from the National Museums of Kenya, 'The process has been to collect and record information from the communities and collect plant materials for food value analysis. The results from the analyses are taken back to the communities, pointing out the more important species to be promoted ...'. Along with demonstration gardens and exhibits at agricultural fairs, one of the main tools in this extension program is the quarterly IFPP Newsletter. It contains descriptions of edible plants, accounts of the dietary habits of local ethnic groups, nutritional information on traditional foods, recipes of favorite dishes and assessments of the successes and failures of the program. Although it reaches neighboring African countries and colleagues in other continents, its primary objective is to recover local knowledge about edible plants and promote their use in rural areas where malnutrition is a serious problem.

8.5.3 Popular publications, exhibits and workshops

Throughout much of the last century, ethnobotanists have published the results of their research in academic journals and books, far from the grasp of the local people with whom they have studied. One of the primary goals of many applied projects is to return the results of previous and current research to rural communities and to society as a whole.

Because scholarly publications are written in a technical style, often in a foreign language, they are not appropriate for distribution to the general public. For this reason, researchers are experimenting with a variety of educational materials, including popular booklets on plant resources and exhibits of medicinal plants as well as with other approaches, such as community botanical gardens, newsletters and indigenous theater. The goal is to provide an accurate record of local ecological knowledge along with relevant findings drawn from taxonomic, phytochemical and other scientific studies. These initiatives face the difficult task of producing a hybrid knowledge, often in written form, while encouraging people to continue experimenting and observing in order to learn more about the natural environment.

As part of a broad program in the early 1980s to promote the survival and development of indigenous culture in the Sierra Norte of Oaxaca, a number of Chinantec, Mixe and Zapotec cultural promoters working with the National Department of Education prepared exhibits of useful plants collected in their communities. They mounted the plants on large pieces of posterboard and included labels that described local uses and names. The exhibits were displayed during community religious and civic festivals, which typically draw people from surrounding settlements. Everyone who visited the exhibit was invited to correct the information or write additional names and uses on a piece of paper that was placed with the specimen. These notes were incorporated into a new label before the exhibit was again displayed. Several popular publications were written by the cultural promoters, drawing upon the results of the surveys and exhibits they realized in their communities.

Participants in the Ethnobotany Task of the Conservation Commission of the Northern Territory, Australia are focusing on promoting traditional Aboriginal plant use through a variety of materials [121, 146]. With the participation of aboriginal people, they have developed plant walks in the Darwin Botanic Garden, posters and popular books. A booklet on Mudburra ethnobotany, for example, contains descriptions of 99 species of useful plants, many of which are illustrated to facilitate identification in the field. The inspiration for the booklets came from the aboriginal people themselves. As the participants in the project wrote:

Due to changes in traditional methods of 'passing on' knowledge (via ceremonies, field based knowledge exchange from senior elders, food and medicine gathering trips and so on) senior Mudburra elders wished to record their traditional knowledge in a contemporary format more suited to contemporary learning patterns. It is intended that this booklet will allow younger Mudburra people to learn their traditional culture in their contemporary school curriculum, whilst at the same time providing a permanent record of Mudburra plant knowledge.... A similar desire has been expressed by many elders from communities and clan groups across the northern areas of the Northern Territory.

Returning the results of surveys and studies of medicinal plants is a permanent activity of TRAMIL, a work group whose approach is partly described in Chapters 3 and 4 [55]. As part of a program called TRAMIL difusión (TRADIF), participants hold workshops in the communities in which they have worked. These events typically bring together students and adults to discuss the proper use of local medicinal herbs. They also provide the opportunity to distribute copies of any popular booklets that have resulted from the study. For example, TRAMIL members from Honduras held a workshop in May of 1991 in the community of Arizona, Atlántida, where they had previously completed 120 medicinal plant surveys in 1990 [147]. Nearly 150 students of various ages took part in events in which they learned about uses and names of local plants, with emphasis on the preparation of remedies and the potential toxicity of some species. A separate meeting was held for 18 adults, who discussed the results of the survey carried out in the community. The *Manual de Plantas Medicinales TRAMIL*, which contains information culled from the community surveys and other sources, was donated to the schools and to the adults that participated in the workshops.

8.6 Use of protected areas

8.6.1 Resource use in protected areas

The regulations of many protected areas in the world strictly prohibit hunting, gathering of plant resources or other productive activities within park or reserve borders. This policy applies to all local people, even those who have traditionally had access to the area and possibly practised sustainable ways of harvesting natural products from the wildlands. They are forced to continue their activities in buffer areas around the park, which may lack some resources and where they may face competition from enterprises which may be ecologically destructive, such as certain types of logging and agriculture.

Through their studies of the sustainability of productive activities, ethnobotanists can act as mediators between local people and park managers, recommending ways in which traditional lifestyles can be compatible with biodiversity conservation. Tony Cunningham, whose work has been cited in various parts of this manual, has demonstrated how a focus on specific issues of resource harvesting can serve to change conservation practice on a national level. As part of an evaluation of parks and people in Bwindi Impenetrable Forest in Uganda – one of two protected area that contains the famed mountain gorillas of Africa – he worked with local resource users and researchers to determine the sustainability of various traditional harvesting activities. Although some of these endeavors were clearly damaging the integrity of the protected area, other practices (such as the collection of honey from stingless wasps) were benign and resulted in a more secure livelihood for local people. By allowing these practices to continue in ways which could be regulated, park managers improved interactions with surrounding communities, increasing local people's involvement in the protection of the Bwindi forest. The compatibility of this approach with national conservation objectives has led the Ugandan government to consider similar accommodations in other parks.

8.6.2 Searching for new products

Richard Evans Schultes, an economic botanist from the United States, has dedicated his life to the discovery of useful plants. From his early exploration of the Sierra Norte of Oaxaca in the 1930s to the many decades of research he later dedicated to the Amazon Basin, he has explored the local use and commercial potential of hundreds of species of tropical plants. His recent book *The Healing Forest*, co-authored with ethnopharmacologist Robert Raffauf, describes nearly 1500 species of medicinal and toxic plants of Northwest Amazonia [148].

The classic goal of economic botany he has pursued – the discovery of new products that can contribute to national development and the well-being of humanity – is undergoing a revival of interest in many developing countries around the world [149]. Javier Caballero, ethnobotanist from the Botanical Garden of the National Autonomous University of Mexico, has asserted that making inventories of useful plants should be high on the list of scientific and technological priorities in all countries of Latin America [150]. Together with colleagues from the Botanical Garden, he is developing a computerized database of all wild and semi-cultivated edible plants of Mexico. The resulting information can be used not only to promote the traditional use of plants but also to detect species which could become important crops on a regional, national or international basis.

Another element in the quest for novel useful plants is the documentation and promotion of local varieties of crop plants. The participants in one notable initiative, Native Seeds/SEARCH, have made over 1200 collections of landraces and wild relatives of crops which are adapted to arid areas in northwestern Mexico and the American Southwest. They cultivate the seeds and redistribute them to local farmers, urban gardeners, researchers and seed banks. Membership of the organization is free to Native Americans, who are also able to obtain any seeds without charge.

Gary Nabhan, one of the founders of Native Seeds/SEARCH, has suggested that some of these local varieties of edible plants could become new crops for small farmers in marginal lands, where rainfall or soil fertility is limited. They are also a source of genetic diversity that may be used in breeding programs to improve the yield, nutritional quality and disease resistance of established crops.

Ethnobotanists should be aware of market conditions before they encourage communities or indigenous organizations to invest their time and resources in new or expanded extractive activities. Although projects of forest product commercialization start with good intentions, they can often end by creating additional economic hardships for local people if carried out in a naive way. Jason Clay [151], who has extensive experience in marketing rainforest products in Europe and the United States, has shared the major lessons he learned in his first years of operation (Box 8.2).

8.6.3 Arts and crafts promotion

When artisans produce baskets, textiles, wood carvings and many other hand-crafted articles, they are at once affirming their cultural identity and guaranteeing an income that supplements their earnings from other productive activities. In a methodological guide to the study of crafts published by UNESCO and the African Cultural Institute, Jocelyne Etienne-Nugue suggests that any promotion of local

Box 8.2 Jason Clay's general principles for developing markets for non-timber forest products

(1) **Start with what is already on the market.** Marketing efforts should focus initially on products which already have markets. The substantial production of these goods can be redirected to new customers or the processing of novel products.

(2) **Diversify products and reduce dependence on a few products.** The diversification of products being sold is absolutely essential to the overall viability of extractivism. In choosing new priority products, attention should be paid to the value generated per unit of labor, as well as to how well the product complements the subsistence base and other productive activities of local people.

(3) **Diversify the markets for raw and processed forest products.** In order to reduce the overall risk to producer groups, attempts should be made to diversify the number and type of end-users of each non-timber forest product.

(4) **Add value locally.** Examine the marketing mechanisms of each product to determine the best ways to capture the value that is added as it leaves the source, including transporting it further into the market, bypassing other traders and processing the commodity locally.

(5) **Capture value that is added further from the source.** All markets in the world add value to products with every transaction or transfer, as well as with increased scarcity or distance from the source of production. Producers, or their supporters, should work to capture some of the value that is added to each step in the commercial system until the product reaches the final consumer.

(6) **Proposed solutions must equal the scope of the problems.** The destruction of rain forests and the displacement of rain forest peoples is proceeding at unprecedented rates. It is not wise to see as solutions any hand-picked or pet projects that are highly subsidized and therefore not reproducible at the scale necessary to solve the problems of many communities.

(7) **No single forest group can provide enough commodities for even a small company in North America or Europe.** The production of a single community or small indigenous group is not sufficient to meet the needs of even a small company and much less those of a large company.

(8) **Controlling a larger market share of a commodity allows considerable influence over the entire market.** This provides the possibility of lobbying for the return of value added and profits to forest residents' organizations or those organizations that work closely with them. Close coordination between all groups is required if these goals are to be achieved.

(9) **Make a decent profit in the marketplace, not a killing.** Do not expect to make a large amount of money in the marketplace – if it were so simple, many more people would be doing it. The goal is to add value locally and increase overall income.

(10) **The markets in North America and Europe are for saving the rain forests rather than for forest people.** It is a sad fact that the awareness of rain forest issues outside of tropical countries is limited to a concern for plants and animals. Few potential customers of such products even know that the forests are occupied by local people.

(11) **Certification of environmental sustainability is key.** Because the growing green market in the United States and Europe is primarily for conservation, the sale of commodities must be linked with systems to ensure that the quantity of products being taken from the forest does not destroy the very resource that consumers are spending money to protect.

artifacts begin by gathering data not only on the handcrafted objects but also on the artisans [104]. Additional notes can be made on the volume of the production and how the merchandise is commercialized. She asserts that careful documentation can yield practical benefits for communities, including attracting support from development agencies, integrating handicrafts into national economic strategies and raising the status and productivity of craft workers.

Because the manufacture of these items is in part dependent on a reliable harvest of forest products, ethnobotanists can promote these artistic traditions by ensuring that all plant and animal materials are being gathered in a sustainable way. In addition, they can explore ways in which handcrafted items can be modified to meet the demands of regional and international markets, particularly keeping in mind the tastes of tourists and art collectors.

In an initiative funded by the Inter-American Foundation, members of a Zapotec community of the Sierra Norte of Oaxaca have explored the use of local dye plants to increase the attractiveness of local crafts. The many weavers in the community traditionally use undyed wool or cotton, producing textiles that are grey, white or black. Additional objects are woven from multicolored agave fibers dyed with chemical substances. Because tourists increasingly demand colorful objects produced with natural dyes, cultural promoters in the community organized a workshop on techniques of dying wool, cotton and agave fibers with local plants. Alejandro de Avila, a Mexican researcher with particular expertise in the textiles and useful plants of southern Mexico, arranged for the participation of a dyer from the neighboring state of Chiapas and a specialist on the marketing of handicrafts. The nearly 20 women from three communities who attended the course produced a large assortment of colors from a few common species found near the community.

8.6.4 Ecotourism

A growing number of tourists are getting off the beaten track to visit sites of exceptional natural beauty. This trend can favor biodiversity conservation – when the tourist activities are well regulated – because visitors contribute revenue for park maintenance and often lend political support for local and national efforts to protect natural areas. These travelers, intrigued by popular reports of ethnobotanical research, are increasingly curious about local people's knowledge of the forest. Their growing appreciation of traditional lifestyles can bolster public backing for policies which favor strengthening the cultural identity and rights of indigenous people. Ethnobotanists can play a key role in designing interpretive programs that bring accurate portrayals of traditional ecological knowledge to tourists, while minimizing the impact of outsiders' visits to indigenous areas.

Naturalists of Kinabalu Park – a 735 km^2 protected area in Sabah, Malaysia – have been incorporating local knowledge of the environment into their environmental education program for a number of years. Of the approximately 200 000 people who visit each year, more than 20 000 climb to the top of Mount Kinabalu, the 4101 m peak that is the centerpiece of the park. They are accompanied by bilingual Dusun guides, many of whom are knowledgeable about the plants and animals found along the summit trail. Ecotourists who do not climb the mountain will soon have another way of discovering Dusun ethnobotany. A newly inaugurated natural history museum and research center in park headquarters will contain exhibits on the use of medicinal plants, rattan palms and other forest products. The information for the interpretive exhibits will be drawn in large part from the Kinabalu Ethnobotany Project, a park-sponsored initiative in which local ethnobotanists are documenting traditional ecological knowledge of plants in their communities.

8.6.5 Health care

Herbal remedies and traditional medical practitioners play an important role in the health care of millions of people in developing countries. In some countries, a large percentage of the population does not have access to Western medical care. Health care can be intimately linked to conservation and ethnobotany, because many medicinal plants are found in habitats endangered by current land use.

A project which integrates ethnobotany, health care and biodiversity conservation is being carried out in the Manongarivo Special Reserve, which covers some 350 km^2 of forested lands in northwest Madagascar. Nat Quansah, a Ghanaian botanist and WWF's Madagascar Plants Officer, has been directing the initiative since its inception in 1989. He is coordinating two teams – one in the field and one in the laboratory – which are documenting traditional medical practices while seeking to improve the availability of health care for local people. The field team consists of local plant specialists who have an intimate knowledge of the flora, university-trained ethnobotanists who carry out surveys of medicinal plant use and medical doctors who diagnose diseases and assess the efficacy of medicinal plants while providing modern medicine to complement the local remedies. The laboratory team is made up of pharmacologists who test the pharmacological activity of medicinal plants and seek to standardize the dosages used in local communities. There is a constant exchange of information between the two teams, allowing local people to evaluate the results of pharmacological tests and laboratory personnel to interpret their own findings in the light of traditional use of the plant remedies.

The central goal of this initiative is to legitimize and promote traditional ways of curing and to make people aware that continued access to plant medicines is dependent on forest conservation. At the same time, it is recognized that there are weaknesses in traditional medical practices. The way forward is seen as integrated health care, in which local medical specialists and western doctors work together to take the best of both traditions, giving preference to the local tradition, other things being equal. Integrated health care is seen as preferable to the present arrangement (common in tropical countries) in which local people have to make a decision on whether to visit traditional or western doctors. At the back of their minds, they are often thinking, 'if this doesn't work, I'll try the other'. The problem is that, by the time the inadequacies of treatment are realized, the patient's condition is likely to have deteriorated considerably. It is thought that the integrated health care approach could have advantages in many countries, based on economic, cultural and environmental criteria.

8.7 The local perspective on ethnobotanical research

Ethnobotanists who come from industrialized countries or from urban centers of developing countries are received in diverse ways by rural communities. The first outsiders are often welcomed uncritically and allowed to work without impediment in the community. The local authorities passively accept the researchers, not necessarily conscious of what benefits or misfortunes the study may eventually bring. In other areas, where the local people have been victims of exploitation by foreigners and urban-dwellers, academics may be prohibited from carrying out field studies. The leaders of still other communities, cognizant of the potential payoffs and pitfalls, agree to collaborate only after ensuring that the research will be carried out in a participatory way and that the results will be used for the benefit of the community.

A community's acceptance of outside researchers often changes as political consciousness grows. In many rural areas, there is increasing discontent with researchers who do not support fundamental political and social change and who provide only limited benefits to local people. As Jason Clay has summarized for tropical biology in general [152]:

> Local individuals or groups who agree to help with surveys of flora and/or fauna rarely receive adequate financial remuneration - unlike the scientists

who work in the area. Furthermore, even though indigenous groups might prove to be the best defenders of the genetic variability that the researchers are attempting to discover and eventually protect, few of the outsiders who collect materials and data from the local population assist the indigenous groups in fighting for land rights.

As part of a larger political movement, local people in many countries have begun to question their participation, whether as informants or research collaborators, in scientific studies carried out in their communities. Many express concerns that they are simply the objects of other people's studies. They do not have access to research results, which are often published in foreign languages and locked away in distant libraries.

Indigenous people's organizations are beginning to address these concerns, adding their voice to the calls for ethical and equitable relationships between all participants in research projects. At a meeting held in Penang, Malaysia in 1992, representatives of indigenous groups produced a charter that advocated new policies regarding scientific investigation, development projects and commercial enterprises that involve their communities. In Article 45 of the charter, they demand that '... all investigations in our territories should be carried out with our consent and under joint control and guidance according to mutual agreement; including provisions for training, publication and support for indigenous institutions necessary to achieve such control'.

The controversy over what benefits local people are to derive from research is taking yet another turn with the current debate over the definition and protection of intellectual property rights. Darrell Posey, an American anthropologist who has long worked in the Brazilian Amazon, has written [52]:

There are no provisions anywhere for the protection of knowledge rights of native peoples. Despite repeated pleas from indigenous leaders that their traditional culture and knowledge systems are being exploited without just compensation, little action has been taken by legal, professional, environmental, non-governmental, governmental – or even human rights – groups to secure intellectual property rights (IPR) for native peoples...

There is considerable opposition to IPR for native peoples – even among ethnobiologists and anthropologists – who fear that they will have to drastically change their 'lifestyles'. Income from published dissertations and other books, slides, magazine articles, phonograph records, films and videos – all will have to include a percentage of the profits to the native 'subjects'. It will probably become normal that such 'rights' be negotiated with native peoples prior to the undertaking of initial fieldwork. This kind of behavior has never been considered as part of the 'professional ethic' of scientific research, but certainly will become so in the near future.

The fight to establish just compensation for intellectual property rights will be a long and difficult process, taking place in many national and international political arenas. In the meantime, what guidelines can be set up for ethnoecological research that will provide for an equitable exchange of knowledge? This topic is being widely debated and several proposals have been advanced. The Declaration of Belem (Box 8.3), product of the First International Congress of Ethnobiology held in 1988, calls for just compensation and legal defense of indigenous peoples' knowledge. The preliminary draft of guidelines of the Society for Economic Botany, formulated by an ethics committee and submitted to the membership for approval in 1990, sets out guidelines for appropriate conduct by fieldworkers. The Panama City Declaration, issued at the end of a workshop held in 1991 on the sustainable harvest and marketing of plant products from tropical rain forests, makes a call to 'hasten the expansion of viable markets guided by ecological and cultural sensitivity and to return economic benefits to rain forest communities in time to halt and ultimately reverse, the deforestation of the tropics'.

Box 8.3 The Declaration of Belem

As ethnobiologists, we are alarmed that:

SINCE

- tropical forests and other fragile ecosystems are disappearing
- many species, both plant and animal, are threatened with extinction
- indigenous cultures around the world are being disrupted and destroyed

and

GIVEN that

- economic, agricultural and health conditions of people are dependent on these resources
- native peoples have been stewards of 99% of the world's genetic resources
- there is an inextricable link between cultural and biological diversity

THEREFORE, as members of the International Society of Ethnobiology, we strongly urge action as follows:

1. Henceforth, a substantial proportion of development aid be devoted to efforts aimed at ethnobiological inventory, conservation and management programs.

2. Mechanisms be established by which indigenous specialists are recognized as proper authorities and are consulted in all programs affecting them, their resources and their environments.

---

3. All other inalienable human rights be recognized and guaranteed, including linguistic identity.

4. Procedures be developed to compensate native peoples for the utilization of their knowledge and their biological resources.

5. Educational programs be implemented to alert the global community to the value of ethnobiological knowledge for human well-being.

6. All medical programs include the recognition of and respect for traditional healers, and the incorporation of traditional health practices that improve the level of health of these populations.

7. Ethnobiologists make available the results of their research to the native peoples with whom they have worked, especially including dissemination in the native language.

8. Exchange of information be promoted among indigenous and peasant peoples regarding conservation, management and sustained utilization of resources.

Other proposals have been forwarded by collectors of seeds, tubers and other samples of **plant germplasm**, which is the reproductive or vegetative material used to propagate plants. The Commission on Plant Genetic Resources (CPGR), part of the Food and Agriculture Organization of the United Nations (FAO), has produced a 'Draft International Code of Conduct for Plant Germplasm Collecting and Transfer' [153]. The Commission urges individual countries to propose their own guidelines and, following this advice, the National Germplasm Resources Laboratory of the United States has released a 'Code of Conduct for Foreign Plant Explorations' [154].

The World Wide Fund for Nature has recently published a discussion paper entitled 'Ethics, Ethnobiological Research and Biodiversity', which focuses on the ethical responsibilities of ethnobotanists as they interact with local communities, national governments, commercial enterprises, research institutes and other parties interested in traditional ecological knowledge [53]. The Manila Declaration, published in 1992 by a UNESCO-sponsored network of Asian botanists and pharmacognosists, proposes ethical guidelines for the use of Asian biological resources [54]. In an appendix to the declaration, specific advice is given to foreign collectors of biological samples (Box 8.4).

These various guidelines, declarations and codes of conduct address the behavior of three broad groups: foreign scientists, host country scientists and local people. An acceptable agreement on how to conduct research must be worked out between all three, respecting the rights and obligations of each. The suggestions presented in Boxes 8.5 and 8.6, culled from various declarations and codes of conduct, can help guide ethnobotanical research.

Box 8.4 Ethical advice for foreign collectors of biological samples

The Manila Declaration contains an appendix which gives concrete advice to foreign collectors of biological samples. The following is a paraphrase of various points to consider:

1. Arrange to work with local scientists and institutes.

2. Respect regulations of the host country – enter the country on a research visa, not a tourist visa and observe rules for export of biological specimens, quarantine, CITES and other national and international regulations.

3. Obtain official permission for all collecting in National Parks or protected areas.

4. Ascertain whether tools and supplies used in scientific work and which are difficult to obtain in the host country can be contributed.

5. When applying for a travel study grant, include equal travel expenses for local counterparts and an amount to cover the cost of processing museum specimens or other costs of the visit to the host institute.

6. Leave a complete set of adequately labelled duplicates with the institute before departing the country.

7. Ensure that types of new species described as a result of the research are deposited in the National Museum or Herbarium of the country of origin.

8. Inform the institute in the country of origin where duplicate specimens are to be deposited.

9. Do not exploit the natural resources of the host country by removing high-value biological products; for example, by collecting without prior permission plants with potential horticultural, medicinal, cultural or other economic value.

10. Obtain a list of rare and endangered species of the country in order to avoid collecting these species without special permission.

11. Collect no more material than is strictly necessary; collect cuttings or seeds rather than uprooting whole plants; whenever possible, collect subsections rather than whole organisms of marine species.

12. Leave photographs or slides for the host institute.

13. Inform the host institute or appropriate organization of the whereabouts of any rare or endangered species that are found.

14. Send copies of research reports and publications to collaborators and host institutes.

15. Acknowledge collaborators and host institutes in research reports and publications.

16. Collect and identify reference voucher specimens for all biological products to be exported.

Box 8.5 Guidelines for conducting ethnobotanical studies with local people, communities and indigenous organizations

1. **Represent and conduct yourself honestly.** Local collaborators must be told your complete institutional affiliation, be it a university, company, governmental agency or non-profit group. Introduce yourself formally to the traditional authorities and leaders of indigenous organizations. This may include presentation of official letters of introduction giving your full name, as well as your local and permanent addresses. The letter should be on official letterhead and signed by an official of your institution. Treat local authorities with the same respect that you would show to a mayor or similar public official in your own city or country.

During your discussion of what the research will entail, explain what you already know about the local environment and what you have learned from others in the region. Explore the ways in which your work may be used for the welfare of the community. Do not attempt to cajole, coerce or trick your informants into revealing data that they consider to be secret. Do not conduct research under false pretense, such as using your activities to influence surreptitiously the religion, culture or economy of local people.

2. **Reveal your source of funding.** The source of funds for the fieldwork should be made explicit, including all obligations required by the funding agency. When possible, a copy of the proposal and grant contract should be presented to collaborating scientists, traditional authorities or local counterparts. It is particularly important to reveal any intentions to commercialize local knowledge or natural resources, by yourself or by your sponsor.

3. **Explain your objectives and methods.** Before beginning your research, visit the traditional authorities of the community to ask for their permission and cooperation. Describe what you plan to do and what you are seeking to learn. Whenever possible, present your project in a public forum such as a town meeting, or to an interested group of citizens organized as a community health committee or similar body. Be prepared to repeat your objectives and methods at any time, particularly when you meet local people in the field or visit them in their homes. Demonstrate what you are accomplishing by sharing with local people the pressed plants, transcribed lists of plant names, photographs of landscapes and other visual aids developed in the course of the study.

4. **Collect botanical samples carefully.** Select your plant material with conservation of local genetic diversity in mind. Do not deplete plant populations, particularly when collecting large samples for phytochemical analysis or for germplasm conservation. Ask permission before gathering any plants which are cultivated or managed. Be aware that stands of plants that

appear to be wild may in fact be protected by local people. Ensure that the locality in which you are collecting has no special spiritual significance which entails protection of plant populations by local people. When in doubt, enquire before you collect.

When making collections of agricultural produce, show appreciation for the seeds, fruits and tubers that farmers give you. To you it may be just another germplasm accession, but as any proud gardener knows, harvesting exemplary produce implies hard labor and great pride.

5. **Return the results**. Let the community and indigenous organizations be aware of all manuscripts and published works that result from the research, providing copies of those that local people might wish to consult. Search for ways to provide technical information in a usable form by, for example, translating materials into languages spoken in the region, convening workshops and exhibits to present the results of the study and offering data that aid the execution of development projects.

6. **Give credit to collaborators**. In each project, you should discuss the advantages and disadvantages of disclosure with your local counterparts and make an informed decision that follows their wishes. When local people give their assent, publish the name and location of the communities that hosted the study and identify the key informants and research assistants. Co-author works with local people or support their efforts to publish their perspectives on the research and their analyses of the results.

In areas where retribution is taken against rural people who try to better their standard of living, gain title to their land or fight for basic human rights, you may endanger your collaborators' lives by publishing or even mentioning their names. Even under less dramatic conditions, some local people may prefer to go unnamed for personal or cultural reasons. Given the lack of strict guidelines for compensating the intellectual property rights of indigenous people, some communities may ask that local knowledge about resources be kept secret. If requested, respect local people's right to anonymity or their desire to keep some information confidential.

7. **Compensate your counterparts**. Compensate those who offer their time, skills or expertise. Establish a form of payment (cash, gifts, exchange of services) that will not disrupt local lifestyles or economic patterns and that will not encourage people to give improvised data. If materials gathered during the course of your research prove to be marketable, work with the community and commercial enterprises to negotiate a contract or other formal agreement which ensures equitable distribution of the profits.

8. **Build local infrastructure**. Strengthen the ability of local people, communities and indigenous organizations to carry out their own ethnobotanical research. Contribute plant specimens to regional and community herbaria,

give books to research centers, leave extra sets of dryers, presses and other collecting equipment. Encourage international and governmental authorities to consider local people as expert consultants and to confer with them when assessing the social and environmental impact of development programs. Promote the involvement of traditional healers into local health programs. Search for ways of creating an exchange of plant knowledge between local people and reinforcing networks of communication between communities. Help community leaders to contact foundations and aid them in preparing proposals for grants and technical support.

9. **Continue your efforts back home.** Lobby development and research agencies, encouraging them to commit resources to programs aimed at inventorying, conserving and managing natural resources. Contact your lawmakers, urging them to introduce and support legislation that guarantees the intellectual property rights of indigenous people, providing a legal infrastructure for the protection of cultural knowledge and natural areas. Query business leaders, inquiring how they plan to return profits to local people and to contribute to the safeguarding of the environment. Work towards educating the general public about the global impact of the threats facing the indigenous peoples and natural areas of the world.

Keep in mind these following recommendations are not hard and fast rules that must be applied dogmatically in all cases. They must be adapted according to the economic, social and political conditions of each country, the abilities of the researchers and the needs of host institutions and local people. The best way to decide on how to behave appropriately is to discuss your research plans with all participants in the project and to come to a common agreement on how to proceed. Take a participatory approach to the work, modifying the research design, fieldwork, analysis and application of the results according to the recommendations of collaborating scientists and local participants. Remember that establishing a good working relationship is an end in itself. Do not be rushed in your attempt to collect and publish data, but adapt yourself to the rhythm, the concerns and the customs of your hosts.

8.7.1 Local people's guidelines for collaboration

In some cases, counterpart guidelines are being established by indigenous organizations to govern the behavior of local people interested in collaborating with academics in ethnobotanical studies. These codes of conduct include recommendations about sharing benefits of the research equally among community members and ensuring the accuracy of data given to outside researchers.

A notable example is the code of conduct proposed by members of PEMASKY – the Spanish acronym for the Project for the Study and Management of Wildlife

Box 8.6 Guidelines for conducting ethnobotanical studies with national scientists

Codes of conduct are being adopted by governmental agencies and research institutions in developing countries that recommend how national and foreign scientists can best work together. These codes often stipulate how to obtain official permission to conduct research and the number of duplicate specimens that are to be left in the host country. In accordance with these codes of conduct, the following points should be kept in mind if you are a foreign scientist involved in a joint project with scientists of the host country:

1. **Consult with colleagues well in advance of beginning research.** You must ensure that your proposed research will not duplicate studies that are already underway. Prepare for your fieldwork by requesting and reading scientific works, published or not, which have been written by colleagues in the host country. As part of preliminary arrangements, decide how scientific collections (such as herbarium specimens) will be distributed and how data (including computerized databases and field notes) will be shared with academic institutions in the host country. Early dialogue about the research enhances the opportunities of interacting with students who may wish to join the research team. Allow sufficient time to apply for any official permits that may be required.

2. **Share the costs of research equitably.** Project budgets from foundations and governmental agencies in the developed world often go far in developing countries, where research funds are usually quite limited. Offer to pay for wear-and-tear on facilities and equipment in the host country, including maintenance of field vehicles, consumption of expendable supplies and other items. Bring as much of your own equipment as possible and provide colleagues with extra quantities of supplies that are difficult to find in the host country.

3. **Seek written approval from the host country for all research activities.** Obtain all official permits from the appropriate governmental agency or national research institution before beginning your research. If requested by any of the participants in the work, set out in writing the rights and responsibilities of visiting and national scientists who will be collaborating on the research. When you conduct fieldwork, carry several sets of photocopies of permits and letters of introduction, which you can leave with the appropriate authorities. Remember that you are obliged to follow all national and international laws that regulate collection and transport of biological specimens and you are required to comply with all arrangements discussed with your colleagues from the host country.

4. Respect the advice of host scientists about local travel and customs. Rely on the recommendations of scientists who travel widely and carry out fieldwork in different parts of their country when you plan your itinerary. Staying out of regions where there is unrest can avoid trouble for everyone – yourself, your colleagues and the local people.

It is prudent to follow your colleagues' behavior when passing government checkpoints and asking the assistance of local authorities. If your colleagues have experience in working with a local ethnic minority, listen to their advice about asking for permission to work in the area and following local customs.

5. Follow through on promises of cooperation after leaving the field. If possible, write a preliminary report of findings before you leave the host country and distribute it to colleagues, institutions and local communities that played a major role in your work. Upon returning home, arrange for permanent labeling and distribution of biological specimens in timely fashion. Correspond with foreign colleagues, particularly with students who are seeking guidance and with co-workers who are continuing to carry out fieldwork. Fulfil any agreements to write up research results and discoveries jointly. Send reprints of publications that have resulted from your visit to all collaborating institutions, colleagues, indigenous organizations and communities.

Areas of Kuna Yala – which focuses on the 60 000 ha Kuna Yala reserve in Panama. One of the primary elements of the code is that a Kuna counterpart accompany each visiting researcher during field studies [155]. Through this arrangement, the local people learn the basics of biological science and acquire the tools and techniques to eventually carry out their own studies. The visiting researchers pay the Kuna co-researchers, providing them with the opportunity to take time away from their subsistence and commercial activities in order to explore the natural resources of the reserve. In order to guide the joint work, PEMASKY produced a booklet on scientific cooperation which describes how to apply for permission to work in the reserve, which localities are open to scientific studies and how to collect plant specimens and monitor animal populations in accordance with the wishes of the local people [156].

Agreements which outline the rights and obligations of all participants in ethnobotanical research will guarantee that we all work together to promote the welfare of local people, equitable distribution of resources, sustainable development and protection of natural areas even as we carry out scientifically rigorous research. An important lesson from previous experiences is to work with local people from the inception of the project and to pay attention to their suggestions and priorities throughout the course of the work. The proposals for international cooperation of the Coordinating Body for Indigenous Peoples' Organizations of the Amazon Basin, presented in Box 8.7, give a broad overview of the issues about which many rural communities voice concern [157].

Box 8.7 Concrete proposals from indigenous peoples for international cooperation

The following statement, made by the Coordinating Body for Indigenous Peoples' Organizations of the Amazon Basin (COICA), has been reprinted in publications of various institutions, including the International Work Group for Indigenous Affairs (IWGIA) and the World Bank. Although focused specifically on the indigenous peoples of the Amazon, it gives a broad perspective on the ways that outsiders can contribute to the well-being of rural communities. Ethnobotanists have the capacity to participate in many of these initiatives, particularly the ones I have emphasized in *italics*. Many of these issues are addressed by the conservation and community development projects discussed in this chapter.

For many decades now, most of our people have been experimenting with ways to participate in the encroaching market economies of our respective countries, while trying to survive as peoples intimately linked to the Amazonian forest. We have done this despite the hostility shown to us by the frontier society and despite the fact that, within the context of the market economy, we are desperately poor. For these reasons, we have organized ourselves in new ways, and developed and managed a variety of small programs to improve our health, education and economy. The following is a brief listing which suggests the kinds of programs which we are currently undertaking or wish to undertake. It is these small-scale, locally controlled initiatives which should be the cornerstone of future Amazonian development.

Programs for territorial demarcation and defense, including *research on territorial composition, land use patterns, soil and forest classifications*; demarcation of territories; titling and registration of territories; training of paralegals, topographers; relocation of settlers and miners squatting on indigenous territories; recuperation of lands illegally taken; *the establishment of complementary forest reserves, wildlife reserves, national parks and joint programs to manage them.*

Programs for resource management, including *research on land use capabilities, soil quality, inventories of flora, fauna and mineral reserves, indigenous management practices; training in research methodology; projects for managing forests through sustainable harvesting practices; projects for improving the productivity of rubber, Brazil nut and other extractive activities; projects of recovering lands and resources devastated by conquest and colonization.*

Programs to strengthen material self-sufficiency, including *research on traditional crops, foods gathered from the forest, farming practices, hunting and fishing technologies; projects for improving productivity, stability and diversity of traditional farming systems;* projects to introduce or improve small animal husbandry; *projects to manage food resources found in the forest; projects to replenish and manage flora for housing, clothing and utensils.*

Programs for economic development, including *projects for industrialization on a small scale of products extracted from the forest; projects to adapt traditional artisan products to market demands;* establishment of community marketing channels; establishment of community-controlled transportation systems; projects to improve productivity of agriculture and animal husbandry where directed at the market.

Programs for maintaining a healthy community, including *research on traditional healing practices, traditional medicines, health problems common to indigenous communities; projects to strengthen traditional health practices;* projects to improve drinking water, nutrition and sanitary conditions where deficient; community-controlled health systems including primary care, diagnostic services and stores of basic medicines; education and training for health care personnel.

Programs for bilingual and intercultural education, including *research on the linguistics of Amazonian languages,* on pedagogies relevant to our situations and cultures; *training for indigenous teachers, linguists and pedagogues; preparation of education materials.*

Programs to defend our rights as peoples, including research on reported violations of indigenous peoples' rights, on Indian customary law; training of indigenous lawyers and paralegals; recourse to top legal advice when necessary; participation in fora promoting the rights of indigenous peoples; campaigns to end slavery, captive communities, debt peonage and forced labor among indigenous peoples; campaigns against forced removals or relocations of indigenous peoples.

Programs for research and documentation, including the coordination and systematization of information relevant to the programs of indigenous peoples within their organization; *establishment of libraries and research centers in the service of indigenous peoples and others who seek new models for Amazonian development.*

Programs for strengthening and communicating our voice, including systems which allow easy communication among indigenous communities and organizations; participation in local, regional, national and international fora where decisions are made which affect our well-being; visits and exchange of experiences among indigenous communities, organizations and programs.

8.8 The path ahead

Over the coming years, colleagues from various fields will continue to focus on the link between ethnobotany, conservation and community development [123]. As conservationists become increasingly interested in understanding plant resource management, there will be additional emphasis on forming multidisciplinary and multicultural teams of researchers who can examine local plant use from various perspectives.

Spurred on by the urgency of this conservation issue, there will be a tendency to support rapid participatory ethnobotanical inventories, followed by detailed studies on selected resources. Attention will be given to posing hypotheses about the link between resource use and conservation as well as to developing empirical methods to test these ideas.

Particular support will be given to local ethnobotanical promoters and in-country scientists who collaborate in the design of resource use studies, because they are in a unique position to apply the results to community development and nature conservation [158, 159]. Collaborators in local communities will not simply be the beneficiaries of these initiatives, but rather full partners in the process. They will participate in the design and implementation of the research as well as the application of the results. In addition to receiving monetary returns, they will expect assistance in analyzing and reinforcing their traditional culture, including testing the efficacy of local herbal remedies, measuring the sustainability of forest management practices and designing ways of ensuring that knowledge is passed along from one generation to the next.

Conservation organizations – and the general public which supports them – will be increasingly perceptive in their assessment of the successes and failures of ethnobotanical exploration. From the perspective of conservation and community development, the way ahead for ethnobotanists is to follow a path of participatory research guided by explicit research agreements and contracts that define the rights and obligations of all participants at each stage of the project. The potential of ethnobotany to contribute to conservation and community development has been well publicized but the world is now waiting for results.

Bibliography

The ethnobotanical literature has been growing swiftly over the last 100 years. This bibliography, a sampling of the range of sources that ethnobotanists consult, is by no means exhaustive. My hope is that it will encourage readers to expand their knowledge of ethnobotany well beyond the limits of this manual. Because of limitations of space, many excellent references have been omitted only because they were not specifically mentioned in the text.

Many important documents exist only as unpublished papers or as manuscripts intended for local distribution. In this bibliography, I put the emphasis on published references that are widely available in libraries throughout the world, but I include some difficult-to-find works that are specifically mentioned in the text.

References

1. Malhotra, K.C., Poffenberger, M., Bhattacharya, A. and Dev, D. (1992) Rapid appraisal methodology trials in Southwest Bengal: assessing natural forest regeneration patterns and non-wood forest product harvesting practices. *Forests, Trees and People Newsletter* **15/16**, 18–25.
2. Phillips, O. (1993) The potential for harvesting fruits in tropical rainforests: new data from Amazonian Peru. *Biodiversity and Conservation* **2**, 18–38.
3. Phillips, O. and Gentry, A.H. (1993a) The useful plants of Tambopata, Peru. I. Statistical hypotheses tests with a new quantitative technique. *Economic Botany* **47**(1), 15–32.
4. Phillips, O. and Gentry, A.H. (1993b) The useful plants of Tambopata, Peru. II. Additional hypothesis testing in quantitative ethnobotany. *Economic Botany* **47**(1), 33–43.
5. Guerrero, R.O. and Robledo, I. (1990) Actividades biologicas de plantas del Bosque Nacional del Caribe (Puerto Rico). *Brenesia* **33**, 19–36.
6. Berlin, B., Breedlove, D.E., Laughlin, R.M., and Raven, P. (1973) Lexical retention and cultural significance in Tzeltal–Tzotzil ethnobotany, in *Meaning in Mayan Languages*, (ed. Edmondson, M.S.), pp. 143–64, Mouton, The Hague.
7. Berlin, B., Breedlove, D.E. and Raven, P.H. (1974) *Principles of Tzeltal Plant Classification*, Academic Press, New York and London.
8. Lewis, W.H. and Elvin-Lewis, M.P. (1994) Basic, quantitative and experimental research phases of future ethnobotany with reference to the medicinal plants of South America in *Ethnobotany and the Search for New Drugs*, CIBA Foundation Symposium 185, pp. 60–76. Wiley, Chichester.
9. Theis, J. and Grady, H.M. (1991) *Participatory Rapid Appraisal for Community Development*. International Institute for Environment and Development, London.

10. World Resources Institute (1991) *Participatory Rural Appraisal Handbook.* Natural Resources Management Support Series No. 1, World Resources Institute, Washington DC.
11. Weber, W.A. (1982) Mnemonic three-letter acronyms for the families of vascular plants: a device for more effective herbarium curation. *Taxon* **31**, 74–88.
12. Hayslett, Jr H.T. (1968) *Statistics Made Simple.* Doubleday, New York.
13. Kachigan, S.M. (1986) *Statistical Analysis, an Interdisciplinary Introduction to Univariate and Multivariate Methods.* Radius Press, New York.
14. Nabhan, G.P. (1985) *Gathering the Desert.* The University of Arizona Press, Tucson, Arizona.
15. Nabhan, G.P. (1989) *Enduring Seeds: Native American Agriculture and Wild Plant Conservation.* North Point Press, San Francisco, California.
16. Davis, W. (1991) Towards a new synthesis in ethnobotany. In *Las Plantas y el Hombre* (eds Rios, M. and Borgtoft Pedersen, H.) Ediciones Abya-Yala, Quito, Ecuador.
17. Hunn, E.S. (1992) The use of sound recordings as voucher specimens and stimulus materials in ethnozoological research. *Journal of Ethnobiology* **12**(2), 187–202.
18. Chazdon, R.L. (1988) Conservation-conscious collecting: concerns and guidelines. *Principes* **32**(1), 13–17.
19. Cremers, G. and Hoff, M. (1990) *Realisation d'un Herbier Tropical.* Institut Francais de Recherche scientifique pour le Developpement en Cooperation. Cayenne, French Guiana.
20. Liesner, R. (n.d.) Field techniques used by Missouri Botanical Garden. Unpublished manuscript.
21. Johnson, D. (ed.) (1991) *Palms for Human Needs in Asia: Palm Utilization and Conservation in India, Indonesia, Malaysia and the Philippines.* A.A. Balkema, Rotterdam.
22. Balick, M.J. (1989) Collection and preparation of palm specimens, in *Floristic Inventory of Tropical Countries*, (eds Campbell, D.G. and Hammond, H.D.), pp 482–83. New York Botanical Garden, New York.
23. Dransfield, J. (1981) The biology of the Asiatic rattans in relation to the rattan trade and conservation, in *The Biological Aspects of Rare Plant Conservation*, (ed. Synge, H.), pp. 179–86. Wiley, London.
24. Dransfield, J. (1988) Prospects for rattan cultivation. *Advances in Economic Botany* **6**, 190–200.
25. Forman, L. and Bridson, D. eds (1991) *The Herbarium Handbook.* Royal Botanic Gardens, Kew.
26. Ter Welle, B.J.H. (1989) Collection and preparation of bark and wood samples, in *Floristic Inventory of Tropical Countries*, (eds Campbell, D. and Hammond, H.D.), New York Botanical Garden, New York.
27. Chapman, C.D.G. (1989) Collection strategies for the wild relatives of crops, in *The Use of Plant Genetic Resources.* (eds Brown, A.H.D., Frankel, O.H., Marshall, D.R. and Williams, J.T.) Cambridge University Press, Cambridge.
28. Hays, T.E. (1976) An empirical method for the identification of covert categories in ethnobiology. *American Ethnologist* **3**, 489–507.
29. Jain, S.K. (1977) *A Handbook of Field and Herbarium Methods.* Today & Tomorrow Printers & Publishers, New Delhi, India.
30. Campbell, D.G. and Hammond, H.D. (eds) (1989) *Floristic Inventory of Tropical Countries.* The New York Botanical Garden, New York.
31. Evans, W.C. (1989) *Trease and Evans' Pharmacognosy*, 13th edn. Bailliere Tindall, London.

32. Brett, J. (n.d.) Why is a plant medicinal?: medicinal plant selection criteria. Unpublished manuscript.
33. Balick, M.J. (1990) Ethnobotany and the identification of therapeutic agents from the rain forest. In *Bioactive Compounds from Plants*, (eds Chadwick, D.J. and Marsh, J.), Ciba Foundation Symposium No. 154, Wiley, Chichester, UK.
34. Elisabetsky, E. and Castilhos, Z.C. (1990) Plants used as analgesics by Amazonian Caboclos as a basis for selecting plants for investigation. *International Journal of Crude Drug Research* **28**, 309–20.
35. Browner, C.H., Ortiz de Montellano, B.R. and Rubel, A.J. (1988) A methodology for cross-cultural ethnomedical research. *Current Anthropology*, **29**, 681–702.
36. Trotter, R.T. (1986) Informant consensus: a new approach for identifying potentially effective medicinal plants, in *Plants in Indigenous Medicine and Diet: Biobehavorial Approaches*, (ed. Etkin, N.L.), pp. 91–112. Redgrave, Bedford Hills, New York.
37. Johns, T., Kokwaro, J.O. and Kimanani, E.K. (1990) Herbal remedies of the Luo of Siaya District, Kenya: establishing quantitative criteria for consensus. *Economic Botany* **44**(3), 369–81.
38. Booth, S., Bressani, R. and Johns, T. (199?) Nutrient content of selected indigenous leafy vegetables consumed by the Kekchi people of Alta Verapaz, Guatemala. *Journal of Food Composition and Analysis*.
39. Johns, T. (1990) *With Bitter Herbs They Shall Eat It: Chemical Ecology and the Origins of Human Diet and Medicine*. The University of Arizona Press, Tucson, Arizona.
40. Omar, S. (1987) Phytochemical expedition to the Danum Valley conservation area, Ulu Segama, Sabah, in *Proceedings of the 3rd Meeting of the Natural Products Research Group, Kota Kinabalu, Sabah, Malaysia*, Universiti Kebangsaan, Malaysia.
41. Jain, S.K. (ed.) (1989) *Methods and Approaches in Ethnobotany*. Society of Ethnobotany, Lucknow, India.
42. Mehrotra, B.N. (1989) Collection and processing of plants for biological screening. In *Methods and Approaches in Ethnobotany*, (ed. Jain, S.K.), pp. 25–37. Society of Ethnobotany, Lucknow, India.
43. Hall, P. and Bawa, K. (1993) Methods to assess the impact of extraction of non-timber tropical forest products on plant populations. *Economic Botany*, **47**(3), 234–47.
44. Balick, M.J., Rivier, L. and Plowman, T. (1982) The effects of field preservation on alkaloid content of fresh coca leaves (*Erythroxylum* spp.). *Journal of Ethnopharmacology*, **6**, 287–91.
45. Cooper-Driver, G.A. and Balick, M.J. (1978) Effects of field preservation on the flavonoid content of *Jessenia bataua*. *Botanical Museum Leaflets* (Harvard University) **26**, 257–65.
46. Toia, R. (1988) The phytochemical survey in relation to natural product chemistry. *Proceedings of the First Meeting of the Natural Products Research Group*. Universiti Kebangsaan Malaysia, Bangi.
47. Elisabetsky, E. (1986) New directions in ethnopharmacology. *Journal of Ethnobiology* **6**, 121–28.
48. Elisabetsky, E. and de Moraes, J.A.R. (1990) Ethnopharmacology: A technological development strategy. In *Ethnobiology: Implications and Applications*. Proceedings of the First Congress of Ethnobiology. Vol. 2, pp. 111–18. Museu Paraense Emilio Goeldi, Belem, Brazil.
49. Croom, Jr, E.M. (1983) Documenting and evaluating herbal remedies. *Economic Botany* **37**, 13–27.
50. Boom, B.M. (1990) Ethics in ethnopharmacology, in *Ethnobiology: Implications and*

Applications. Proceedings of the First International Congress of Ethnobiology, Vol. 2, 147–53. Museu Emilio Goeldi, Belem, Brazil.

51. Posey, D. (1990a) Intellectual property rights and just compensation for indigenous knowledge. *Anthropology Today*, 6(4), 13–16.

52. Posey, D.A. (1990b) Intellectual property rights: what is the position of ethnobiology? *Journal of Ethnobiology* 10, 93–98.

53. Cunningham, A. (1993) *Ethics, Ethnobiological Research and Biodiversity.* WWF, Gland, Switzerland.

54. Anon (1992) *The Manila Declaration Concerning the Ethical Utilization of Asian Biological Resources.* UNESCO Regional Network for the Chemistry of Natural Products in Southeast Asia, Selangor, Malaysia.

55. Robineau, L. (ed.) (1991) *Towards a Caribbean Pharmacopoeia, Scientific Research and Popular Use of Medicinal Plants in the Caribbean,* Enda-caribe, Santo Domingo.

56. Kottak, C.P. (1991) *Anthropology, the Exploration of Human Diversity.* 5th edn, McGraw-Hill, New York.

57. Brim, J.A. and Sapin, D.H. (1974) *Research Design in Anthropology: Paradigms and Pragmatics in the Testing of Hypotheses,* Holt, Rinehart and Winston, New York.

58. Weller, S.C. and Romney, A.K. (1988) *Systematic Data Collection.* Qualitative Research Methods Series 10, Sage Publications, Newbury Park, UK.

59. Lipp, F.J. (1989) Methods for ethnopharmacological fieldwork. *Journal of Ethnopharmacology,* 25, 139–50.

60. Aumeeruddy, Y. (n.d.) Management of a Sacred Forest in the Kerinci Valley, Central Sumatra; an example of conservation of cultural and biological diversity. Unpublished manuscript.

61. Aumeeruddy, Y. (1993) Agroforêts et aires de forêts protégées. Representations et practiques agroforestières paysannes en périphérie du Parc National Kerinci Seblat, Sumatra, Indonésie. Doctoral Thesis, Université de Montpelier II, France.

62. Levi-Strauss, C. (1966) *The Savage Mind,* University of Chicago Press, Chicago, Illinois.

63. Smith-Bowan, E. (1954) *Return to Laughter.* Harper Brothers, New York.

64. Conklin, H.C. (1954) The Relation of Hanunoo Culture to the Plant World. Unpublished PhD dissertation.

65. De Avila, Alejandro (n.d.) Plants in contemporary Mixtec ritual: *Juncus, Nicotiana, Datura* and *Solandra.* Unpublished manuscript.

66. Boster, J. (1980) How exceptions prove the rule: An analysis of informant disagreement in Aguaruna Manioc classification. Unpublished PhD dissertation, University of California, Berkeley.

67. Boster, J.S. (1984) Classification, cultivation and selection of Aguaruna cultivars of *Manihot esculenta* (Euphorbiaceae), in *Ethnobotany in the Neotropics* (eds Prance, G.T. and Kallunki, J.A.), *Advances in Economic Botany,* Vol. 1, pp. 34–47.

68. Banack, S.A. (1991) Plants and Polynesian voyaging, in *Islands, Plants and Polynesians: An Introduction to Polynesian Ethnobotany* (eds Cox, P.A. and Banack, S.A.), Dioscorides Press, Portland, Oregon.

69. Lebot, V. (1991) Kava (*Piper methysticum* Forst. f) The Polynesian dispersal of an oceanian plant in *Islands, Plants and Polynesians: An Introduction to Polynesian Ethnobotany* (eds Cox, P.A. and Banack, S.A.), Dioscorides Press, Portland, Oregon.

70. Barley, N. (1983) *The Innocent Anthropologist, Notes from a Mud Hut,* Henry Holt, New York.

71. Friedberg, C. (1968) Les methodes d'enquete en ethnobotanique. *Journal d'Agriculture Tropical et Botanique Appliquee* **15**(7–8), 297–324.

72. Friedberg, C. (1976) Questionnaires ethnobotaniques, in *Enquete et Description des Langues a Tradition Orale* (eds Bouquiaux, L. and Thomas, J.M.C.), Vol. 3. SELAF, Paris.

73. Giron, L.M., Freire, V., Alonzo, A. and Caceres, A. (1991) Ethnobotanical survey of the medicinal flora used by the Caribs of Guatemala. *Journal of Ethnopharmacology* **34**, 173–87.

74. Hunn, E.S. (1990) *Nch'i-Wana, The Big River: Mid-Columbia Indians and Their Land.* University of Washington Press, Seattle, Washington.

75. Anderson, J.M. and Ingram, J.S.I. (1993) *Tropical Soil Biology and Fertility: A Handbook of Methods*, 2nd edn, CAB International, Wallingford, Oxon, UK.

76. Foth, H.D. and Turk, L.M. (1972) *Fundamentals of Soil Science*, 5th edn, Wiley, New York.

77. McIntosh, R.P. (1985) *The Background of Ecology: Concept and Theory.* Cambridge University Press, Cambridge.

78. Walujo, E.B. (1988) Les Ecosystèmes Domestiqués par l'Homme dans l'Ancien Royaume Insana-Timor (Indonésie). Unpublished doctoral thesis, L'Université Paris VI (Pierre et Marie Curie), France.

79. Malhotra, K.C. (1993) People biodiversity and regenerating tropical sal (*Shorea robusta*) forests in West Bengal, India, in *Tropical Forests, People and Food. Biocultural Interactions and Applications to Development*, (eds Hladik, C.M. *et al.*), UNESCO, Paris, and Parthenon Publishing, Carnforth, UK.

80. Touber, L., Smaling, E.M.A., Andriesse, W. and Hakkeling, R.T.A. (1989) *Inventory and Evaluation of Tropical Forest Land, Guidelines for a Common Methodology.* TROPENBOS Technical Series 4. The Tropenbos Foundation, Ede, The Netherlands.

81. Boom, B. (1989) Use of plant resources by the Chacobo. *Advances in Economic Botany*, **7**, 78–96.

82. Campbell, D.G. (1989) Quantitative inventory of tropical forests, in *Floristic Inventory of Tropical Countries* (eds Campbell, D.G. and Hammond, H.D.), The New York Botanical Garden, New York.

83. Prance, G.T. (1991) What is ethnobotany today? *Journal of Ethnopharmacology*, **32**, 209–16.

84. Salick, J. (1992) Amuesha Forest Use and Management: an integration of indigenous use and natural forest management, in *Conservation of Neotropical Forests: Working from Traditional Resource Use* (eds Redford, K.H. and Padoch, C.), pp. 305–32, Columbia University Press, New York.

85. Dallmeier, F. (1992) *Long-term Monitoring of Biological Diversity in Tropical Forest Areas, Methods for Establishment and Inventory of Permanent Plots.* MAB Digest 11. UNESCO, Paris.

86. Howard, P.C. (1991) *Nature Conservation in Uganda's Tropical Forest Reserves.* IUCN, Gland, Switzerland.

87. Toledo, V.M., Batis, A.I., Becerra, R., Martinez, E. and Ramos, C.H. (1992) Products from the tropical rainforests of Mexico: an ethnoecological approach, in *Sustainable Harvest and Marketing of Rain Forest Products* (eds Plotkin, M. and Famolare, L.), pp. 99–109, Island Press, Washington DC.

88. Rico-Gray, V., Garcia-Franco, J.G., Chemas, A. and Sima, P. (1990) Species composition, similarity and structure of Mayan homegardens in Tixpeual and Tixcacaltuyub, Yucatan, Mexico. *Economic Botany* **44**(4), 470–87.

89. Alcorn, J.B. (1989) An economic analysis of Huastec Mayan forest management, in *Fragile Lands of Latin America, Strategies for Sustainable Development* (ed. Browder, J.O.), Westview Press, Boulder, Colorado.

90. Balick, M.J. and Mendelsohn, R. (1992) Assessing the economic value of traditional medicines from tropical rain forests. *Conservation Biology* **6**, 128–30.

91. Godoy, R.A., Lubowski, R. and Markandaya, A. (1993) A method for the economic valuation of non-timber forest products. *Economic Botany* **47**(3), 220–33.

92. Padoch, C. (1987) The economic importance and marketing of forest and fallow products in the Iquitos region. *Advances in Economic Botany* **5**, 74–89.

93. Padoch, C. (1988) Aguaje (*Mauritia flexuosa* L.f.) in the economy of Iquitos, Peru. *Advances in Economic Botany* **6**, 214–24.

94. Peters, C.M., Gentry, A.H. and Mendelsohn, R.O. (1989) Valuation of an Amazonian rainforest. *Nature* **339**, 655–56.

95. Pinedo-Vasquez, M., Zarin, D. and Jipp, P. (1992) Economic returns from forest conversion in the Peruvian Amazon. *Ecological Economics* **6**, 163–73.

96. Nepstad, D.C. and Schwartzman, S. (eds) (1992) *Non-timber Products from Tropical Forests: Evaluation of a Conservation and Development Strategy*. Advances in Economic Botany, Vol. 9, The New York Botanical Garden, New York.

97. Plotkin, M. and Famolare, L. (1992) *Sustainable Harvest and Marketing of Rain Forest Products*. Island Press, Washington DC.

98. Swanson, T.M. and Barbier, E.B. (1992) *Economics for the Wilds: Wildlife, Wildlands, Diversity and Development*. Earthscan Publications, London.

99. Munasinghe, M. (1992) Biodiversity protection policy: environmental valuation and distribution issues. *Ambio* **21**, 227–36.

100. Ehrlich, P.R. and Ehrlich, A.H. (1992) The value of biodiversity. *Ambio* **21**, 219–26.

101. Kant, S. and Mehta, N.G. (1993) A forest-based tribal economy: a case study of Motisingloti village. *Forest, Trees and People Newsletter* **20**, 34–39.

102. Cunningham, A. and Mbenkum, F.T. (1993) *Sustainability of Harvesting of Prunus africana Bark in Cameroon: A Medicinal Plant in International Trade*. People and Plants working paper 2, UNESCO, Paris.

103. Malhotra, K.C., Deb, D., Dutta, M. *et al.* (1992) Role of non-timber forest produce in village economy: a household survey in Jamboni Range, Midnapore District, West Bengal. Working paper of the Indian Institute of Bio-Social Research and Development.

104. Etienne-Nugue, J. (1990) *Crafts, Methodological Guide to the Collection of Data*. UNESCO, Paris.

105. Martin, G. (1992) Searching for plants in peasant marketplaces. In *Sustainable Harvest and Marketing of Rain Forest Products* (eds Plotkin, M. and Famolare, L.), pp. 212–23, Island Press, Washington DC.

106. Bye, Jr R.A. and Linares, E. (1983) The role of plants found in the Mexican markets and their importance in ethnobotanical studies. *Journal of Ethnobiology* **3**(1) 1–13.

107. Pike, K.L. (1971) *Phonemics: A Technique for Reducing Languages to Writing*. University of Michigan Press, Ann Arbor, Michigan.

108. Ladefoged, P. (1982) *A Course in Phonetics*, 2nd edn, Harcourt Brace Jovanovich, New York.

109. Felger, R.S. and Moser, M.B. (1985) *People of the Desert and Sea, Ethnobotany of the Seri Indians*. University of Arizona Press, Tucson, Arizona.

110. Balee, W. (1989) Nomenclatural patterns in Ka'apor ethnobotany. *Journal of Ethnobiology* **9**(1), 1–24.

111. Carroll, J.B. (1956) *Language, Thought and Reality: Selected Writings of Benjamin Lee Whorf*, Massachusetts Institute of Technology, Cambridge, Massachusetts.

112. Balee, W. and Moore, D. (1991) Similarity and variation in plant names in five Tupi-Guarani languages (Eastern Amazonia). Bulletin of the Florida Museum of Natural History. *Biological Sciences* **35**(4), 209–62.

113. Berlin, B. (1992) *Ethnobiological Classification: Principles of Categorization of Plants and Animals in Traditional Societies*. Princeton University Press, Princeton, New Jersey.

114. Brown, C.H. (1984) *Language and Living Things: Uniformities in Folk Classification and Naming*. Rutgers University Press, New Brunswick, New Jersey.

115. Conklin, H. (1962) Lexicographical treatment of folk taxonomies. *International Journal of American Linguistics* **28**, 119–41.

116. Berlin, B. (1973) Folk systematics in relation to biological classification and nomenclature. *Annual Review of Ecology and Systematics*, **4**, 259–71.

117. Frake, C.O. (1980) *Language and Cultural Description*. Stanford University Press, Stanford, California.

118. Hunn, E. (1977) *Tzeltal Folk Zoology: The Classification of Discontinuities in Nature*. Academic Press, New York.

119. Bennett, B.C. (1992) Plants and people of the Amazonian rainforests: The role of ethnobotany in sustainable development. *Bioscience*, **42**(8), 599–607.

120. McNeely, J.A., Miller, K.R., Reid, W.V., Mittermeier, R.A. and Werner, T.B. (1990) *Conserving the World's Biological Diversity*. IUCN, Gland.

121. Williams, N.M. and Baines, G. (1993) *Traditional Ecological Knowledge: Wisdom for Sustainable Development*. Centre for Resource and Environmental Studies, Canberra, Australia.

122. Hladik, C.M., Hladik, A., Linares, O.F. *et al.* (eds) (1993) *Tropical Forests, People and Food: Biocultural Interactions and Applications to Development*. UNESCO, Paris, and Parthenon Publishing, Carnforth, UK.

123. Martin, G. (1994) Conservation and ethnobotanical exploration, in *Ethnobotany and the Search for New Drugs*, CIBA Foundation Symposium 185, Wiley, Chichester.

124. Posey, D.A. and Balee, W. (eds) (1989) *Resource Management in Amazonia: Indigenous and Folk Strategies*. Advances in Economic Botany, vol. 7, The New York Botanical Garden, New York.

125. Posey, D.A. and Overal, W.L. (organizers) (1990) *Ethnobiology: Implications and Applications*. Proceedings of the First International Congress of Ethnobiology (Belem, 1988). Volumes 1 & 2. Museu Paraense Emilio Goeldi, Belem, Brazil.

126. Redford, K.H. and Padoch, C. (eds). (1992) *Conservation of Neotropical Forests: Working from Traditional Resource Use*. Columbia University Press, New York.

127. Alcorn, J.B. (1984) Development policy, forests and peasant farms: reflections on Huastec managed forests' contributions to commercial production and resource conservation. *Economic Botany* **38**, 389–406.

128. Brokensha, D.W., Warren, D.M. and Werner, O. (1978) *Indigenous Knowledge Systems and Development*. University Press of America, Washington DC.

129. Ryan, J.C. (1992) *Life Support: Conserving Biological Diversity*. Worldwatch paper 108. Worldwatch Institute, Washington DC.

130. United States National Research Council (1992) *Conserving Biodiversity: A Research Agenda for Development Agencies*. National Academy Press, Washington DC.

131. World Resources Institute (1992) *Global Biodiversity Strategy*. World Resources Institute, Washington DC.

132. Toledo, V.M. (1982) La Etnobotánica Hoy: Reversión del Conocimiento, Lucha Indígena, y Proyecto Nacional. *Biótica*, **7**(2), 141–50.

133. Chapin, M. (1988) The seduction of models: Chinampa agriculture in Mexico. *Grassroots Development* 12(1), 8–17.
134. Altieri, M. (1987) *Agroecology: Scientific Basis for an Alternative Agriculture*, Westview Press, Boulder, Colorado.
135. Altieri, M.A., Anderson, K. and Merrick, L.C. (1987) Peasant agriculture and conservation of crop and wild plant resources. *Conservation Biology* 1, 49–58.
136. Gliessman, S.R., Garcia, E. and Amador, A. (1981) The ecological basis for the application of traditional agricultural technology in the management of tropical agroecosystems. *Agroecosystems* 7, 173–85.
137. Alcorn, J.B. (1981) Huastec non-crop resource management. *Human Ecology*, 9, 395–417.
138. Brush, S.B. (1986) Genetic diversity and conservation in traditional farming systems. *Journal of Ethnobiology* 1, 109–23.
139. Shiva, V., Anderson, P., Schucking, H. *et al.* (1992) *The Rainforest Harvest: Sustainable Strategies for Saving the Tropical Rainforests?* Friends of the Earth, London.
140. Nabhan, G.P. (1992) Native plant products from the arid Neotropical species: assessing benefits to cultural, environmental and genetic diversity. In *Sustainable Harvest and Marketing of Rain Forest Products* (eds Plotkin, M. and Famolare, L.), pp. 137–40. Island Press, Washington DC.
141. Homma, A.K.O. (1992) The dynamics of extraction in Amazonia: a historical perspective, in *Non-timber Products from Tropical Forests: Evaluation of a Conservation and Development Strategy* (eds Nepstad, D.C. and Swartzman, S.), Advances in Economic Botany, Vol. 9, pp. 23–31.
142. Lescure, J.P., Emperaire, L., Pinton, F. and Renault-Lescure, O. (1992) Nontimber forest products and extractive reserves in the middle Rio Negro region, Brazil, in *Sustainable Harvest and Marketing of Rain Forest Products* (eds Plotkin, M. and Famolare, L.), pp. 151–57, Island Press, Washington DC.
143. Lescure, J.P. and Pinton, F. (1993) Extractivism: a controversial use of the tropical ecosystem, in *Tropical Forests, People and Food. Biocultural Interactions and Applications to Development*. (eds Hladik, C.M. *et al.*), UNESCO, Paris, and Parthenon Publishing, Carnforth, UK.
144. Nabhan, G.P. (1985b) Native crop diversity in Aridoamerica: conservation of regional gene pools. *Economic Botany* 39, 387–99.
145. Meilleur, B.A. The ethnobotanical garden and tropical plant conservation, in *Tropical Botanic Gardens: Conservation and Development*. BGCS/IUCN (eds.) Academic Press, London and New York.
146. Wightman, G., Dixon, D., Williams, L. and Dalywaters, I. (1992) *Mudburra Ethnobotany: Aboriginal Plant Use from Kulumindini (Elliott) Northern Australia*, Northern Territory Botanical Bulletin No. 14, Conservation Commission of the Northern Territory, Darwin, Australia.
147. House, P., Lagos-Witte, S. and Torres, C. (1989) *Manual Popular de 50 Plantas Medicinales de Honduras*. Tegucigalpa, Honduras: CONS-H, CIIR, UNAH.
148. Schultes, R.E. and Raffauf, R.F. (1990) The Healing Forest, Medicinal and Toxic Plants of the Northwest Amazonia. Dioscorides Press, Portland, Oregon.
149. Reid, W.V., Laird, S.A., Meyer, C.M. *et al.* (eds) (1993) *Biodiversity Prospecting: Using Genetic Resources for Sustainable Development*. World Resources Institute, Washington.
150. Caballero, J. (1987) Etnobotánica y desarrollo: la búsqueda de nuevos recursos vegetales, in *Memorias del Simposio de Etnobotánica del IV Congreso Latioamericano de Botánica* (ed. Toledo, V.M.), pp. 79–96. Instituto Colombiano para el Fomento de la Educación Superior, Bogotá, Colombia.

151. Clay, J. (1992) Some general principles and strategies for developing markets in North America and Europe for nontimber forest products. In *Sustainable Harvest and Marketing of Rain Forest Products* (eds Plotkin, M. and Famolare, L.), Island Press, Washington DC.

152. Clay, J.W. (1988) *Indigenous Peoples and Tropical Forests – Models of Land Use and Management from Latin America.* Cultural Survival Inc., Cambridge, Massachusetts.

153. FAO (1991) *Draft International Code of Conduct for Plant Germplasm Collecting and Transfer.* Commission on Plant Genetic Resources, fourth session. 15–19 April, 1991. FAO, Rome.

154. National Germplasm Resources Laboratory (1990) *Code of Conduct for Foreign Plant Explorations.* National Germplasm Resources Laboratory, Beltsville, Maryland.

155. Chapin, M. (1991) How the Kuna keep the scientists in line. *Cultural Survival Quarterly* 15(3), 17.

156. PEMASKY (1988) Programa de Investigación, Monitoreo y Cooperación Científica: Información para los investigadores. PEMASKY, Panama.

157. Davis, Sheldon H. (1993) *Indigenous Views of Land and the Environment.* World Bank Discussion Paper 188. The World Bank, Washington DC.

158. Berlin, B. (1984) Contributions of Native American Collectors to the Ethnobotany of the Neotropics, in *Ethnobotany in the Neotropics* (eds Prance, G.T. and Kallunki, J.A.), Advances in Economic Botany, Vol. 1, The New Botanical Garden, New York.

159. Wells, M. and Brandon, K. (1992) *People and Parks, Linking Protected Area Management with Local Communities.* The World Bank, Washington DC.

Further reading

Atran, S. (1990) *Cognitive Foundations of Natural History*, Cambridge University Press, Cambridge.

Barrera, A. (1979) La etnobotánica, in *La Etnobotánica: tres puntos de vista y una perspectiva* (ed. Barrera, A.) INIREB, Xalapa, Veracruz.

Bellamy, R. (1993) *Ethnobiology in Tropical Forests, Expedition Field Techniques.* Royal Geographic Society, London.

Castetter, E.F. (1944) The domain of ethnobiology. *American Naturalist* 78, 158–70.

Denslow, J.S. and Padoch, C. (1988) *People of the Tropical Rainforest.* University of California Press, Berkeley, CA.

Ellen, R. and Reason, D. (1979) *Classifications in Their Social Context.* Academic Press, London.

Ford, R.I. (1978) Ethnobotany: historical diversity and synthesis, in The Nature and Status of Ethnobotany (ed. Ford, R.I.), pp. 33–50, Anthropological Papers 67, University of Michigan, Ann Arbor, Michigan.

Godoy, R.A. and Bawa, K.S. (1993) The economic value and sustainable harvest of plants and animals from the tropical forest: Assumptions, hypotheses and methods. *Economic Botany* 47(3), 215–19.

Prance, G.T., Balée, W., Boom, B.M. and Carneiro, R.L. (1987) Quantitative ethnobotany and the case for conservation in Amazonia. *Conservation Biology* 1, 296–310.

Scheps, R. (ed) (1993) *La Science Sauvage: Des Savoirs Populaires aux Ethnosciences.* Editions du Seuil, Paris.

Toledo, V.M. (1991) *El Juego de la Supervivencia: Un Manual para la Investigación Etnoecológica en Latinoamerica.* Consorcio Latinoamericano sobre Agroecología y Desarrollo, Berkeley, California.

Toledo, V.M. (1992) What is ethnoecology? Origins, scope and implications of a rising discipline. *Etnoecológica* 1(1), 5–21.

Index

Page numbers in **bold** refer to words indexed by their main definition only. In general, latin plant names are indexed; vernacular names only where useful.

41 74